Food Ethics

Franz-Theo Gottwald · Hans Werner Ingensiep ·
Marc Meinhardt
Editors

Food Ethics

Editors
Franz-Theo Gottwald
Schweisfurth Foundation
Suedliches Schlossrondrell 1
80638 München
Germany
cthomas@schweisfurth.de

Hans Werner Ingensiep
Institute für Philosophy
University of Duisburg-Essen
Universitätsstr. 12
45117 Essen
Germany
h.w.ingensiep@uni-due.de

Marc Meinhardt
Institute für Philosophy
University of Duisburg-Essen
Universitätsstr. 12
45117 Essen
Germany
marcmeinhardt@gmx.de

ISBN 978-1-4899-8456-2   ISBN 978-1-4419-5765-8 (eBook)
DOI 10.1007/978-1-4419-5765-8
Springer New York Dordrecht Heidelberg London

© Springer Science+Business Media, LLC 2010
Softcover re-print of the Hardcover 1st edition 2010
All rights reserved. This work may not be translated or copied in whole or in part without the written permission of the publisher (Springer Science+Business Media, LLC, 233 Spring Street, New York, NY 10013, USA), except for brief excerpts in connection with reviews or scholarly analysis. Use in connection with any form of information storage and retrieval, electronic adaptation, computer software, or by similar or dissimilar methodology now known or hereafter developed is forbidden.
The use in this publication of trade names, trademarks, service marks, and similar terms, even if they are not identified as such, is not to be taken as an expression of opinion as to whether or not they are subject to proprietary rights.

Printed on acid-free paper

Springer is part of Springer Science+Business Media (www.springer.com)

Contents

1 **Introduction: Food Ethics in a Globalized World – Reality and Utopia** 1
Hans Werner Ingensiep and Marc Meinhardt

Part I Food, Consumers, and Policy

2 **The Ethical Matrix as a Tool in Policy Interventions: The Obesity Crisis** 17
Ben Mepham

3 **Ethical Traceability for Improved Transparency in the Food Chain** 31
Christian Coff

Part II GM-Food Production

4 **Ethics and Genetically Modified Foods** 49
Gary Comstock

5 **Responsible Agro-Food Biotechnology: Precaution as Public Reflexivity and Ongoing Engagement in the Service of Sustainable Development** 67
Marian Deblonde

6 **Precautionary Approaches to Genetically Modified Organisms and the Need for Biosafety Research** 87
Anne Ingeborg Myhr

7 **Biotechnology, Battery Farming and Animal Dignity** 101
Peter Kunzmann

Part III Food, Globalization, and Water

8 **Agricultural Trade and the Human Right to Food: The Case of Small Rice Producers in Ghana, Honduras, and Indonesia** 119
Armin Paasch, Frank Garbers, and Thomas Hirsch

9   **Hunger, Poverty, and Climate Change: Institutional Approaches, a New Business Alliance, and Civil Courage to Live Up to Ethical Standards** 137
    Franz-Theo Gottwald

10 **Food Versus Fuel: Governance Potential for Water Rivalry** 153
    Lena Partzsch and Sara Hughes

11 **Whose Nature – Whose Water? Some Remarks About the History of Ideas, Property and Democracy of Water** 167
    Uta von Winterfeld

12 **Towards a New Architecture of Agricultural Trade in the World Market** 185
    Wolfgang Sachs and Tilman Santarius

13 **Epilogue: The Schweisfurth Foundation – A German Food-Ethics-Platform** 205
    Franz-Theo Gottwald

Index 209

About the Authors

Christian Coff, PhD in food ethics, is an agricultural scientist and was Research Director at the Center for Ethics and Law in Copenhagen. Coff founded the first consumer-supported agriculture organization (CSA) in Denmark in 2001 and published the book "The Taste for Ethics: An Ethic of Food Consumption" (2006).

Gary L. Comstock is Professor for Philosophy at North Carolina State University, Raleigh, N.C, a member of EURSAFE, and was Director of the Research and Professional Ethics Program at NC State University (2002–2007). Comstock organized international projects in the field of bioethics and is editor of the textbook "Life Science Ethics" (2002).

Marian Deblonde, PhD in Applied Philosophy, works for the IST, Flemish Parliament, Brussels, and was senior researcher at IMDO Technology Assessment research group, the Institute for Environment and Sustainable Development of the University of Antwerp. Deblonde has researched and published on central questions in technology risk assessment, sustainable development, participative methods, and political decision making.

Frank Garbers, PhD in Anthropology at the University of Hamburg, is a consultant for agrarian politics and development cooperation. Garbers worked for many years in Guatemala and researched the impact of globalization on peasant economies. He has published articles about the free trade agreement between Central America and the USA and the right to food, agrarian reform, and rural development.

Franz-Theo Gottwald, Honorary Professor for Environmental-, Agrarian and Nutritional Ethics at Humboldt University, Berlin, is Executive Director of the Schweisfurth Foundation, Munich, Germany. Gottwald has published many articles and is co-editor of many books in the field of agrarian culture, management and philosophy.

Thomas Hirsch, geographer and social scientist, has worked on rural development, agricultural trade, human rights, and ecology and was engaged in local and international NGOs in Latin America and Asia. Hirsch was for eight years the Finance Director of the human rights organization FIAN International and works now as policy advisor on Climate Change and Food Security for the Protestant development cooperation agency Bread for the World (Brot für die Welt) in Germany.

Sara Hughes, PhD Environmental Science and Management, Master of Science, Fisheries and Wildlife, works at the University of California, Santa Barbara. Hughes has taught, researched, and published in the field of virtual water, sustainable development, environmental law, and politics.

Hans Werner Ingensiep, PhD biology and Professor for Philosophy, teaches at the Center for Medical Biotechnology at the University of Duisburg-Essen, Germany. He has published several books in the field of biophilosophy, bioethics, and food philosophy for example as co-editor of the anthology "Leben, Töten, Essen – Anthropologische Dimensionen" (2000).

Peter Kunzmann, PhD, philosopher and theologian, is Professor for Bioethics at the Center for Ethics, University of Jena, Germany. Kunzmann worked in a research project on the "dignity" of animals and plants in gene technology and published several articles and books in the field of philosophy and animal ethics for example "Die Würde des Tieres – zwischen Leerformel und Prinzip" (2007).

Marc Meinhardt MA, studied philosophy, political science, and history at the University of Duisburg-Essen. Meinhardt has published and worked on the philosophical and political dimensions of Food Ethics.

Ben Mepham was formerly Director of the Center for Applied Bioethics at the University of Nottingham. Since retirement, he has held honorary professorships in bioethics at the School of Biosciences, University of Nottingham and in the Department of Policy Studies, University of Lincoln. Mepham was a founder member of the UK Government's Biotechnology Commission, the European Society for Agricultural and Food Ethics (EURSAFE), and the Food Ethics Council (of which he was the first Executive Director). In 2008 Mepham published the second edition of his highly acclaimed textbook "Bioethics: An introduction for the biosciences."

Anne Ingeborg Myhr, PhD studied biotechnology and is a scientist at the Norwegian Institute of Gene Ecology in Tromsø. She worked on "Precaution, Context and Sustainability. A Study of How Ethical Values may be Involved in Risk Governance of GMOs." Myhr's present research interests in an interdisciplinary project are DNA vaccines, the elaboration of philosophical perspectives on GMOs, and capacity building in risk assessment and management of GMO use and release in the Third World.

Armin Paasch, Master in history at Cologne University, is the Trade Officer of the German section of the organization FoodFirst Information and Action Network (FIAN), the international Human Rights Organization for the Right to Food. Paasch has conducted various studies on agrarian reforms, about the impact of trade policies on the right to food and about hunger policies.

Lena Partzsch, PhD in political sciences, has worked on environmental policy and is member of the department of economics at the Helmholtz Center for Environmental Research in Germany. Partzsch worked and published on global and new governance modes in environmental and developmental policies, water science, and European Integration.

Wolfgang Sachs, PhD, sociologist, Honorary Professor at the University of Kassel, works at the Wuppertal Institute for Climate, Environment, and Energy, Germany. In the last decades he was a member of many research groups dealing with energy, environment, sustainability, globalization, and developmental politics. Sachs has been active in Greenpeace, Germany, and is a member of the Club of Rome. He has published many articles and books in the wide field of international developmental policy for example "The Development Dictionary: A Guide to Knowledge as Power" (1992).

Tilman Santarius, MA in sociology, anthropology, and economics works for the Heinrich-Böll-Foundation and Germanwatch e.V. and was Senior Research Fellow at the Wuppertal Institute for Climate, Environment, and Energy, Germany. Santarius worked in the area of economic instruments in climate policy, global governance, and issues regarding trade and environment. With Wolfgang Sachs he is co-author of "Fair Future. Limited Resources, Security, and Global Justice" (2007).

Uta von Winterfeld, PhD studied philosophy and political science and is the leader of the research project "Sustainable Production and Consumption" at the Wuppertal Institute for Climate, Environment, and Energy, Germany. Von Winterfeld researched and published in the fields of social relations to nature, feminist criticisms of science, and sustainable employment.

Contributors

Christian Coff University College Sjælland, Ankerhus, Denmark.

Gary Comstock Iowa State University, Raleigh, NC, USA, gcomstock@ncsu.edu

Marian Deblonde Institute Society and Technology – Flemish Parliament, Brussels, Belgium, marin.deblonde@vlaamsparlement.be

Frank Garbers Terre des Hommes, Osnabrück, Germany, info@garbers.info

Franz-Theo Gottwald CEO Schweisfurth Foundation, Munich, Germany, cthomas@schweisfurth.de

Thomas Hirsch Brot für die Welt, Stuttgart, Germany, t.hirsch@brot-fuer-die-welt.de

Sara Hughes Bren School of Environmental Science and Management, University of California, Santa Barbara, CA, USA, shughes@bren.ucsb.edu

Hans Werner Ingensiep Institute for Philosophy, University of Duisburg-Essen, Essen, Germany, h.w.ingensiep@uni-due.de

Peter Kunzmann Ethikzentrum Jena, Jena, Germany, peter.kunzmann@uni-jena.de

Marc Meinhardt Institute for Philosophy, University of Duisburg-Essen, Essen, Germany, marcmeinhardt@gmx.de

Ben Mepham Center for Applied Bioethics, University of Nottingham, Nottingham, UK, ben.mepham@nottingham.ac.uk

Anne Ingeborg Myhr Genøk – Center for Biosafety, Tromsø, Norway, anne.i.myhr@uit.no

Armin Paasch Misereor, Aachen, Germany, armin.paasch@misereor.de

Lena Partzsch Department of Economics, Helmholtz Center for Environmental Research, Leipzig, Germany, lena.partzsch@ufz.de

Wolfgang Sachs Wuppertal Institute, Wuppertal, Germany, wolfgang.sachs@wupperinst.org

Tilman Santarius Heinrich-Böll-Foundation, Berlin, santarius@boell.de

Uta von Winterfeld Wuppertal Institut für Klima, Umwelt, Energie, Wuppertal, Germany, uta.winterfeld@wupperinst.org

Abbreviations

AML	Animal microencephalic lumps
AMPU	Anand Milk Producers Union
AoA	Agreement on Agriculture
BMI	Body Mass Index
BMU	Federal Ministry of Environment (Germany)
BULOG	Badan Urusan Logistic
CA	Consumers Autonomy
CAC	Codex Alimentarius Commission
CAP	Common Agriculture Policy
CARMEL	Cooperativa Asociativa de Campesinos de Transformacion y Servicios Otorena
CBD	Convention on Biological Diversity
CESCR	Committee on Economic, Social, and Cultural Rights
CF	Consumers Fairness
CIF	Council of International Fellowship
COM	Communication of the European Commission on the Precautionary Principle
CW	Consumers Wellbeing
DR-CAFTA	Free Trade Agreement between USA and Central American Countries
EAA	Ecumenical Advocacy Alliance
EED	German Church Development Service
EFSA	European Food Safety Authority
EKAH	Swiss Federal Ethics Committee on Nonhuman Biotechnology
EM	Ethical Matrix
ERS	Economic Research Service
ETO	Extraterritorial Obligations
EU	European Union
EurepGAP	European Good Agricultural Practices
EURSAFE	European Society for Agricultural and Food Ethics
FAO	Food and Agriculture Organization
FAOSTAD	Statistical Database of the United Nations Food and Agriculture Organization

FIAN	FoodFirst Information and Action Network
GATS	General Agreement on Trade in Services
GATT	General Agreement on Tariffs and Trade
GDFL	Grameen Danone Food Ltd.
GMO	Genetically Modified Organism
HACCP	Hazard Analysis of Critical Control Points
HSSFF	High sugar, salt, and fat foods
ICESCR	International Convent on Economic, Social and Cultural Rights
ICFFA	International Commission on the Future of Food and Agriculture
IFOAM	International Federation of Organic Agriculture Movements
IGO	Intergovernmental Organization
IHMA	Honduran Institute of Agricultural Marketing
ILO	International Labor Organization
IMF	International Monetary Fond
IPC	International Policy Council
IPCC	Intergovernmental Panel on Climate Change
IPPC	International Plant Protection Commission
ISO	International Organization for Standardization
MA	Marketers Autonomy
MDG	Millennium Development Goals
MERCUSOR	Mercado Común del Sur
MF	Marketers Fairness
MOFA	Ministry of Food and Agriculture
Mt	Metric tonnes
MW	Marketers Wellbeing
NAFTA	North American Free Trade Agreement
NBC	National Biosafety Commission
NC	Naïve Consequentialism
NEG	Negative
NGO	Non-Governmental Organization
NHS	National Health Service
NTNU	Norges teknisk-naturvitenskapelige universitet
OECD	Organization of Economic Cooperation and Development
OIE	World Organization for Epizootics
PA	Producers Autonomy
PF	Producers Fairness
POM	Policy Objectives Matrix
POS	Positive
PW	Producers Wellbeing
R&D	Research and Development
RSPCA	Royal Society for the Protection of Animals
RSPO	Roundtable on Sustainable Palm Oil
SA	Societies Autonomy
SAP	Structural Adjustment Programme

SF	Societies Fairness
SPM	Specified Principles Matrix
SPS	Agreement on Sanitary and Phytosanitary Measures
SW	Societies Wellbeing
TBT	Agreement to Technical Barriers to Trade
TRIMS	Agreement on Trade-Related Investment Measures
TWN	Third World Network
UNCTAD	United Nations Conference on Trade and Development
UNDP	United Nations Development Program
UNEP	United Nations Environmental Programme
UNESCO	United Nations Educational, Scientific and Cultural Organization
USD	United States Dollars
USDA	United States Food and Drugs Administration
W&H framework	Walker and Harremöes framework
WBGU	Wissenschaftlicher Beirat der Bundesregierung Globale Umweltveränderungen (Germany)
WSSCC	Water Supply and Sanitation Collaborative Council
WSSD	World Summit of Sustainable Development
WTO	World Trade Organization
WWF	World Wide Fund for Nature

Chapter 1
Introduction: Food Ethics in a Globalized World – Reality and Utopia

Hans Werner Ingensiep and Marc Meinhardt

Food Ethics has become an essential part of a globalized world and thus the question "What is Food Ethics?" seems almost superfluous. At the beginning of the twenty-first century about one billion people are starving all over the world, while the industrial nations are locked in debate about the financial crisis or climate change. At the same time there are many interdisciplinary discussions about special problems in the field of applied ethics and in this new and young discipline called Food Ethics. Experts mainly in Western societies discuss everything pertaining to the entire food chain, like obesity, traceability, agro-food biotechnology, transgenic plants, biofuels, and the world trade system. On the other hand, for years non-governmental organizations (NGOs) have been lamenting politicians' lack of interest in creating or using existing ethical tools for solutions geared toward changing this situation. This background motivated us to compile this book and hence give some insights into current discussions and important issues within the field of Food Ethics. The main principles, tools and case studies are presented in this book as well as useful information at national and international levels. Today we are still far from Food Ethics being a regular subject at schools or universities – even though it should be and will be of great interest in the future. However, it is questionable whether Food Ethics should be understood as a homogeneous discipline. Many scientists of diverse disciplines are currently working on Food Ethics, and each of them has chosen a different theoretical or practical approach. Within the scope of Food Ethics the focus of interest is not only on reaching scientific but also economic and/or political goals. In politics or economics ethical matters in particular have to be taken into account. Today, the image of business and management ethics is enhanced when Food Ethics standards are applied.

This book may be useful as a first step and introduction to the field of Food Ethics. It will be helpful to all those interested in generating a higher consciousness of national and global perspectives on Food Ethics topics. The authors who are experts within the Food Ethics community provide insights into important ethical

H.W. Ingensiep (✉)
Institute for Philosophy, University of Duisburg-Essen, Essen, Germany
e-mail: h.w.ingensiep@uni-due.de

problems, proposals for ethical tools and also present engaged theoretical positions or political solutions. The articles focus on the reality of global food problems mainly on two levels. On the first level they discuss current questions of nutrition in specific contexts and important fields, for example the obesity crisis, traceability in the food chain, genetically modified foods, scarcity of water resources, and world hunger. The second level concerns ethical tools, important ideas and suggestions for long-term solutions.

On the other hand it is obvious that we need something like an utopian view concerning global food problems. The word "utopia" leaves a bitter taste in a world of around a billion starving people. It is time to remember that utopias were and still are timeless generators for future societies. Utopia delivers political innovation and ethical impulses for reflection. Therefore, in the second part of this contribution we will review some utopian seeds and dreams by authors stretching from the past to the present. Their contributions enable heuristic comparisons for students and experts and allow constructive criticisms based on real facts. The third part gives a short reflection on the special problem of justice from a modern philosophical point of view. But now we introduce with a summary of the contributions to this book into the wide field of real problems in Food Ethics in a globalized world.

1.1 Food Ethics and Reality

Ben Mepham, professor and director of the Centre for Applied Bioethics at the University of Nottingham, UK, has a great deal of experience of teaching bioethics and has been an investigator in the new field of Food Ethics for more than 30 years. Mepham coined the concept "Food Ethics" and developed an important ethical tool, the so-called "ethical matrix" (Mepham 1996). As a scientist he understands the basis of the biomedical ethics approach of Beauchamp and Childress (2008), and for biosciences he published the textbook "Bioethics: An Introduction for the Biosciences" (Mepham 2005). This author's contribution addresses the issue of how governments of democratic states might seek to assist their citizens in reversing the serious trend towards overweight and obesity. Effective policy interventions need to be implemented on several fronts. This case study demonstrates the possibilities and arguments within the framework of Mepham's influential "ethical matrix" as both a procedural and a substantive tool in such programs, focusing on food production, marketing and consumption.

Christian Coff, agricultural scientist and former research director at the Centre for Ethics and Law in Copenhagen, founded the first consumer-supported agriculture group (CSA) in Denmark in 2001. Consumer concerns include animal welfare, health, environmental issues, and food chain transparency. How can consumers be empowered to influence such ethical concerns? Information is the first priority. But although consumers are constantly bombarded with information on food, details about how foods have actually been produced are hard to find. Other information is superficial, conflicting or partial, thus making informed food choices difficult or impossible. As a consequence food traceability is needed, i.e. information about the history and route of a given food. It is used in the food sector for legal and commercial reasons and has the potential to communicate a more authentic picture of how food is produced.

Gary L. Comstock (2002), professor of philosophy at North Carolina State University, has been focusing on ethical questions in the biosciences and has experience in international projects in the field of bioethics. Comstock explored the central dogma that humans are superior to other life forms, especially to higher animals. In his article Comstock analyzes several ethical arguments concerning food plants as they are discussed in American politics and life science ethics: Is it ethically justifiable to pursue genetically modified crops and foods? There are many specific and general objections, for instance, the statement "biotech is playing God". Comstock tries to give objective answers and discusses the various ethical issues in use by genetic technology in agriculture. Over time he himself changed his mind about the moral acceptability of GM crops. Now Comstock agrees with some heterogeneous arguments: utilitarian considerations and others related to virtue theory and in harmony with the rights principle. In his eyes these three most influential ethical positions (Rights, Utilitarian, and Virtue) support the production of GM plants in a globalized world, and he thinks that it is a basic human right of people in other countries to choose this new technology.

Marian Deblonde, Institute Society & Technology, Flemish Parliament, Brussels, has been engaged in several research projects dealing with technology risk management and political decision making. Deblonde observes that in Europe, agro-food biotechnologies have aroused a lot of controversy in the past four decades of their development. It looks as if many of these issues had been transferred to other new and emergent technologies. Deblonde considers the adequacy of Europe's regulatory reaction by the way it interprets and uses the precautionary principle. One important statement is that a fundamental reinterpretation of this principle is needed and that it should be reconnected to the guiding idea of sustainable development. This implies a collective engagement, compilation of projections for the future, and a continuous learning process for responsible actions. Another central point of her contribution is a criticism of the European risk assessment system.

Anne Ingeborg Myhr studied biotechnology and teaches at the Norwegian Institute of Gene Ecology in Tromsøe. Myhr worked on the principles of precaution and sustainability, and how ethical values are involved in risk governance of genetically modified organisms (GMOs). In her article she shows that many interpretations of the precautionary principle do not provide a basis for adequate decisions on ecological and ethical questions. Myhr disagrees with the opinion that the precautionary principle is not scientifically based and gives reasons for an enlargement of scientific methods. She supports an argumentative application of a concept, which is based on sustainable actions.

Peter Kunzmann, bioethicist and professor at the Centre for Ethics, University of Jena, Germany, has been working on a research project dealing with the "dignity" of animals and plants in gene technology. With respect to the question of the legitimacy of applying biotechnology to animals, he advocates the concept of the "dignity of creatures". Kunzmann analyzes several philosophical problems by comparing them to other basic controversial concepts in animal ethics. The core ethical idea is to defend the plausibility of this concept and to analyze its relation to other important principles in the field like integrity or intrinsic value. Kunzmann shows that a species-proper treatment of animals is contradictory to biotechnological modification of animals. Finally, he gives reasons why the concept of dignity should lead a debate on animal ethics because it includes the guarantee of animal welfare as well as human responsibility for animal treatment.

Armin Paasch is the Trade Officer of the German section of the FoodFirst Information and Action Network (FIAN), the international human rights organization for the right to food. He has conducted various studies on agrarian reforms, about the impact of trade policies on the right to food and about hunger policies. His first co-author, Frank Garbers, anthropologist, worked in Guatemala for several years as a researcher and consultant for farmers' organizations. In this context Garbers was involved in research on the impact of globalization on peasant economies, and he analyzed the consequences of the free trade agreement between Central America and the USA. Other topics of his work are the right to food, agrarian reform and rural development. Thomas Hirsch, geographer and social scientist, has been working since 1992 on rural development, agricultural trade, human rights and ecology. Through his engagements with local as well as with international NGOs he gained field work experience in Latin America and Asia. After 8 years as Finance Director of the human rights organization FIAN International Hirsch is currently working as Policy Advisor on Climate Change and Food Security for the Protestant development cooperation agency "Brot für die Welt" (Bread for the World) in Germany.

Their article summarizes a comprehensive study commissioned by the Ecumenical Advocacy Alliance (EAA), a worldwide network of church organizations, and "Brot für die Welt", and conducted by FIAN, the international human rights organization for the right to food, in cooperation with national and international experts. This study examines the impact of specific rice trade policies on the Human Right to Adequate Food of specific rice-producing communities in Ghana, Honduras and Indonesia. Therefore the authors analyze causal chains, first linking sharp increases of rice imports and hunger, malnutrition and food insecurity, and second showing the link between these import increases and certain trade and agricultural policies. Other possible factors, such as natural disasters, land tenure arrangements or access to infrastructure are given due attention in order to place the influence of trade policies into context. The case studies also include a thorough human rights analysis of the findings, distinguishing between the different obligations and responsibilities of national governments, external states and intergovernmental organizations (IGOs).

Franz-Theo Gottwald, honorary professor for environmental, agricultural and nutritional ethics at Humboldt University, Berlin, has been the Executive Director of the Schweisfurth Foundation in Munich, Germany, for many years, where Gottwald published many books in the fields of agrarian culture, politics, management, religion and philosophy. His contribution introduces the problems of hunger and poverty induced by anthropogenic climate change. Several effects of anthropogenic climate change on agriculture and water resources are discussed. Looking for solutions to the current problems Gottwald discusses conventional and alternative approaches to fighting world hunger and evaluates their effectiveness. He presents the Grameen Danone project as a case-in-point, a project that is successful without development aid, mainly through cooperation between industry and civil society, and operates fairly and sustainably on the basis of the ethical principles of subsidiarity and solidarity.

Lena Partzsch studied political sciences and is a member of the department of economics at the Helmholtz Centre for Environmental Research in Germany. Partzsch worked on global and new governance modes in environmental and developmental policies, water science and European integration together with Sara Hughes, working in Environmental Science and Management at the University of California. Partzsch analyzes the problem of biofuel and food crop production competing for scarce arable land and water. The Mexican "tortilla crisis" in 2007 publicly revealed this dilemma: because of the increasing use of corn for biofuels, the price for corn meal in Mexico increased to three times its price in 2006. Such intersections can create challenges for development of policies – both at the global and national levels – that guarantee affordable and accessible food sources. Further, as water resources continue to be stretched, tradeoffs for consumptive uses will become increasingly common. Virtual water accounting is a useful tool that has been developed to improve our understanding of the way water is used in the production of goods, and particularly how this affects the global distribution of water through trade in these goods. Applying empirical and theoretical understandings of virtual water flows to the assessment and development of biofuel trade and development policies is critical when considering the water costs of their expanded use. These methods can also be expanded in order to apply to food crops. This study assesses the current understanding of global water resources and the impacts of trade on their distribution. One intention is to utilize tools such as virtual water accounting to inform policy makers and to incorporate a wider array of social and environmental goals.

Uta von Winterfeld studied philosophy and political science and is the leader of a research project "Sustainable Production and Consumption" at the Wuppertal Institute for Climate, Environment and Energy, Germany. Von Winterfeld worked on relations of society and nature, feminist criticism of science and sustainable work. Her article includes a historical review of one of the important questions of pre-modern times: How to justify "property" – whether God gave the world, gave nature to all his children, to mankind for their common use? Von Winterfeld tries to give answers and shows how our comprehension of nature is still influenced by Francis Bacon, and that the concept of property developed by John Locke still influences neoliberal theories of private property. Furthermore, she gives insights into Commons and argues for a revision of this concept. This is the background for her criticism of the increasing privatization of common goods like water, and a claim for sustainable and fair handling of global water resources.

The sociologist Wolfgang Sachs, has been part of many research groups and organizations in the last decades dealing with energy, the environment, sustainability, globalization and development politics for instance of the Club of Rome. Sachs published "The Developmental Dictionary. A guide to knowledge as power" (London 1992). Together with his co-author, Tilman Santarius, scientist at the Wuppertal Institute for Climate, Environment and Energy, Germany, and others, Sachs sketches the policy framework for a slow, fair and sustainable trade in agriculture facing globalization and world markets. They all criticize the liberalization of trade that empowers transnational conglomerates by disempowering national politics and they intend to reform the structural adjustment policies, which the World Trade Organization (WTO) has issued during the past 25 years. Once dumping prices and floods of exports have been discouraged, responsible states will be able to set policy frameworks for sustainable family farming. Support of local and regional markets is necessary in order to guarantee food security and sustainability, and the human right to food in a democratic way. Thus, a globalized world needs a new post-WTO architecture for agricultural trade. Perhaps we have now reached the wide field of Utopia, but without Utopia there will be no fair policies in a future globalized world.

1.2 Food Ethics and Utopia

Far from the reality as described and analyzed in the articles of this book, it is always helpful to look back on former utopian material, not only concerning Food Ethics. Thus, here we will look back for some classical food for thought on Utopia. We mentioned some paradigms and positions in philosophy. Today we think or know that it took mankind millions of years to move from "nature" to "culture" during the evolution of the species *homo sapiens sapiens*. In prehistoric times man began as a nomad, hunter or food collector, and later he settled down. Little by little man became a homo agrarian with agricultural cultivation and rural economies. Later again he was organized into high "cultures" like in Egypt, India or China. Very old myths and dreams include parts of an agricultural religious utopia like those illustrated in the ideas of a vegetarian paradise or a "golden age". "Works and Days" was inspired by moral admonitions and written in the eighth century B.C. by the Greek poet Hesiod, who described these "golden times": Human beings "lived like gods without sorrow of heart, remote and free from toil and grief [...] for the fruitful earth unforced bare them fruit abundantly [...] rich in flocks and loved by the blessed gods" (Hesiod 1914, pp. 110–120). These "golden times" disappeared and times of war, hunger or disease followed. After Prometheus' lapse mankind had to work hard on farms, and "agri-culture" as a special kind of "culture" had begun. Hesiod gives advice on farm labour and implements, for example on feeding and relaxation. This is the point, far away from Utopia, where the long path from "cultivation" to "civilization" of ancient Greek history begins.

As an important step towards a new state of technocracy in Western civilization, the next utopian approach seemed to be a very good solution. The philosopher and politician Francis Bacon (1561–1626) in his work "New Atlantis" (1627) presents a vision of a perfect state based upon experimental science. "Salomon's House" is the main science centre and the institution for necessary research facilities for example laboratories and gardens. In this work mankind has to study "nature" by means of observation and experimentation because "knowledge is power". Bacon gives insights into the scientific process and institutions like the later Royal Society, a process leading to modern gene and biotechnology, which helps produce all we need for life: bread, meat, beverages and pills. Bacon: "We also have great variety of composts, and soils, for the making of the earth fruitful. [...] We have also means to make diverse plants rise by mixtures of earths without seeds; and likewise to make diverse new plants, differing from the vulgar; and to make one tree or plant turn into another " (Bacon 1938). Synthetic Biology is not far away from Bacon's Utopia. The questions are often not what is possible, but what is allowed and desirable? Until the seventeenth century utopians like Morus, Bacon and Campanella developed new religious, social, scientific and technical frameworks for an early "globalization", when many new countries were discovered and conquered. Those "modern" utopians were convinced of an exclusive dignity and superiority of man; on the one hand "nature" for them was an object, instrument or resource, and on the other hand "culture" was a field for social activity, property and private interests. One of them, the Italian philosopher Tommaso Campanella (1568–1639), in his book "City of the Sun" (1623) designs a social Utopia on the basis of a form of theocratic communism. We hear about the people in this state "with what foods and drinks they are nourished, and in what way and for how long they live". But at the beginning they had a small ethical problem: "They were unwilling at first to slay animals, because it seemed cruel; but thinking afterwards that it was also cruel to destroy herbs which have a share of sensitive feeling, they saw that they would perish from hunger unless they did an unjustifiable action for the sake of justifiable ones, and so now they all eat meat. Nevertheless, they do not kill willingly useful animals, such as oxen and horses" (Campanella 2004). The people of Utopia feel a collision between their private moral of respect for animals and plants and their obvious needs for nutrition in a civilized society. Campanella thought that plants were a kind of sensitive being. But, even after Descartes' declaration that animals and plants merely are automata without any souls, and after Darwin's explanation that animals and plants are produced by natural selection during evolution, modern writers like Samuel Butler (1835–1902) used similar arguments. In the 27th chapter of his novel "Erewhon" (1872/1901, reprint 1979), which means "nowhere" written backwards, Butler tries to defend the views of an Erewhonian philosopher and botanist concerning the "rights of vegetables". Vegetables are our "cousins" and "animals under another name". But even in Butler's Utopia it is seen as absurd and impossible to justify the killing of animals, not to mention plants. For the anti-Christian and Lamarckian evolutionist Butler respect for plants and animals was ridiculous, but good material for a satirist (see Baranzke et al. 2000). On the other hand, Butler speculates about machines as extensions of the human organism, and that these machines may threaten us by a rival evolution. Today some developments seem to be not far away from this speculation.

Jean-Jacques Rousseau (1712–1778) was not a utopianist. But he brought new ideas into the discussion on the relation between "nature", "culture" and the process of "civilization". With Rousseau we connect the "Social Contract", "liberty", "equality", "fraternity", and the French Revolution starting in 1789. Other ideas of Rousseau are that man is good by nature and corrupted by society and civilization. In his famous "Discourse upon the origin and the foundation of the inequality among mankind" (1755) his questions were: What is the difference between the natural state and the later state of a civilized society with a "high" culture? What is the origin of inequality among mankind? The framework for his answers is complex, a mixture of speculation, construction, history, policy and observations. The "golden" natural state of an autarkic "savage man" for Rousseau seems to be better than the civilized state of a man who has to work, defend private property and earn money for food. There is an important difference between man and animals: "Nature speaks to all animals, and beasts obey her voice. Man feels the same impression, but he at the same time perceives that he is free to resist or to acquiesce; and it is in the consciousness of this liberty, that the spirituality of his soul chiefly appears" (Rousseau 2004). Man is good and free in the natural state and imagined as a kind of isolated hunter. But with agriculture and work the division of labour and the limitation by private property becomes necessary. Competition and inequality develop. According to Rousseau with agriculture the harmony of nature and the natural state of man is destroyed forever. In a civilized society man depends on mutual aid, power, control, rules etc. This led to the division into rich and poor people, slaves and monarchs. Some ideas of Rousseau influenced thinkers as different as Kant or Marx, and the "romanticism" or "back to nature" movements until today. Some of our modern ideas go back to Rousseau, the pre-industrial philosopher. This raises the question whether these approaches are proper solutions for modern social and food problems in a globalized world more and more dominated by industry and technology.

As nineteenth-century analysts and critics of an industrial society had been looking for a new social or scientific-technical utopia, Petr Kropotkin (1842–1921), the famous Russian theorist of anarchism, like Rousseau, believed in the goodness of human nature, even after Spencer's and Darwin's selfish principle of the "survival of the fittest" seemed to justify capitalism and egoism. According to Darwin mankind appeared to be the winner species in the evolution of the earth. Within this species the winners and losers were a product of "natural selection" leading to a hierarchical society without solidarity and equality, but grounded in egoism. Nevertheless Kropotkin wanted to organize this society along the principle of "Mutual Aid" (1902), and without any corrupt authority and without exploitative capitalism. Rather his political and ethical principles support the institution of small "natural" agricultural and industrial communities. In his other book "The Conquest of Bread" (1892, 1906) Kropotkin confesses "that we are Utopians". Kropotkin's clear message is: "Bread, it is bread that the Revolution needs! It has always been the middle-class idea to harangue about 'great principles' – great lies rather! The idea of people will be to provide bread for all. And while middle-class citizens, and workmen infested with middle-class ideas admire their own rhetoric in the 'Talking Shops', and 'practical people' are engaged in endless discussions on forms of government, we, the 'Utopian dreamers' – we shall have to consider the question of daily bread" (Kropotkin 1970, p. 69). What is the solution? "There is only one really practical solution of the problem – bodily to face the great task which awaits us, and instead of trying to patch up a situation which we ourselves have made untenable, to proceed to reorganize production on a new basis" (Kropotkin 1970, p. 74). Does industry help? "Epochs of great industrial discoveries and true progress in industry are precisely those in which the happiness of all was the aim pursued, and in which personal enrichment was least thought of. Great investigators and great inventors aimed, without doubt, at the emancipation of mankind. And if Watt, Stephenson, Jacquard, etc., could have only foreseen what a state of misery their sleepless nights would bring to the workers, they would probably have burned their designs and broken their models" (Kropotkin 1970, p. 266). What did Kropotkin think about agriculture 100 years ago? "The agriculturist has broader ideas today – his conceptions are on far grander scale. He only asks for a fraction of an acre in order to produce sufficient vegetables for a family; and to feed twenty-five horned beasts he needs no more space than he formerly required to feed one; his aim is to make his own soil, to defy seasons and climate, to warm both air and earth around the young plant; to produce, in a word, on 1 acre what he used to crop on 50 acres, and that without any excessive fatigue – by greatly reducing, on the contrary, the total of former labour. He knows that we will be able to feed everybody by giving to the culture of the fields no more time than what each can give with pleasure and joy. This is the present tendency of agriculture" (Kropotkin 1970, p. 269). Solidarity, equality, autarchy were the principles of Kropotkin's political philosophy. His Utopia consists of small communities in agriculture and industry without repression by money and politics. What do we think about this a century later?

"Small is Beautiful. Economics as if People Mattered. 25 Years later" is the well-known slogan of the British economist Ernst Friedrich Schumacher (1911–1977) contained in his famous work (Schumacher 1973). Today his problems are still our problems: "Among material resources, the greatest, unquestionably is the land. Study how a society uses its land, and you can come to pretty reliable conclusions as to what its future will be. The land carries the topsoil, and the topsoil carries an immense variety of living beings including man" (Schumacher 1999, p. 80). Schumacher compared the state of agriculture with modern industries: "In the discussion of the Mansholt Plan, agriculture is generally referred to as one of Europe's 'industries'. The question arises whether agriculture is, in fact, an industry, or whether it might be something essentially different. Not surprisingly, as this is a metaphysical – or meta-economic – question, it is never raised by economists. Now, the fundamental 'principle' of agriculture is that it deals with life, that is to say, with living substances. Its products are the results of processes of life and its means of production is the living soil. A cubic centimetre of fertile soil contains billions of living organisms, the full exploration of which is far beyond the capacities of man. The fundamental 'principle' of modern industry, on the other hand, is that it deals with man-devised, not-living materials. The ideal of industry is the elimination of living substances. Manmade machines work more reliably and more predictably than do such living substances as men. The ideal of industry is to eliminate the living factor, even including the human factor, and turn the productive process over to machines" (Schumacher 1999, p. 87).

Three decades later we have another situation: Bio- and gene technology provide us with useful "living machines" for example transgenic organisms: bacteria, plants and animals. A new industry designs new forms of life and produces "biofacts". Should we remember this "old" statement of Schumacher? "In other words, there can be no doubt that the fundamental 'principles' of agriculture and of industry, far from being compatible with each other, are in opposition. Real life consists of the tensions produced by the incompatibility of opposites, each of which is needed, and just as life would be meaningless without death, agriculture would be meaningless without industry. It remains true, however, that agriculture is primary, whereas industry is secondary, which means that human life can continue without industry, whereas it cannot continue without agriculture" (Schumacher 1999, p. 88). What were the aims in Schumacher's small-is-beautiful-utopia? "On a wider view, however, the land is seen as a priceless asset which it is man's task and happiness 'to dress and to keep'. We can say that man's management of the land must be primarily orientated towards the goals – health, beauty, and permanence. The fourth goal – the only accepted by the experts – productivity, will be attained almost as byproduct. The crude materialist view sees agriculture as 'essentially directed towards food-production'. A wider view sees agriculture as having to fulfil at least three tasks:

– to keep man in touch with living nature, of which he is and remains a highly vulnerable part;
– to humanize and ennoble man's wider habitat; and
– to bring forth the foodstuffs and other materials which are needed for a becoming life" (Schumacher 1999, p. 89).

Again utopia and reality are mixed, if we remember the statements of the American economist Dennis L. Meadows (born 1942) and his colleagues as published in "The Limits to Growth" (Meadows et al. 1972). A study of the Club of Rome published in 2004 stated the now well-known result that "business as usual", like in the last 30 years, will lead to an "overshoot and collapse" of growth in the next three decades: "Still there is enough food, at least in theory, to feed everyone adequately. The total amount of grain produced in the world around the year 2000 could keep eight billion people alive at subsistence level, if it were evenly distributed, not fed to animals, and not lost to pests or allowed to rot between harvest and consumption. Grain constitutes roughly half the world's agricultural output (measured in calories). Add the annual output of tubers, vegetables, fruits, fish, and animal products produced from grazing rather than grain, and there would be enough to give the turn-of-the-millennium population of six billion a varied and healthful diet" (Meadows et al. 2004, p. 57).

Today a new industry of genetic engineering has been established. Some people think that this new industry is the solution to all food problems in a globalized world. Thus, is there more need for utopias and "romanticisms"? Meadows' answer is:

"Why have we not mentioned the promise of genetically modified crops? Because the jury is still out on this technology – indeed the jury is in deep controversy. It is not clear either that genetic engineering is needed to feed the world or that it is sustainable. People are not hungry because there is too little food to buy; they are hungry because they cannot afford to buy food. Producing greater amounts of high-cost food will not help them. And while genetic engineering might increase yields, there are plenty of still-unrealized opportunities to raise yields without genomic interventions that are both high-tech (therefore inaccessible to the ordinary farmer) and ecologically risky. The rush to biotech crops is already producing troubling ecological, agricultural, and consumer backlashes. Everyone could be more than adequately nourished with the amount of food now grown. And more food could be grown. It could be done with much less pollution, on less land, using less fossil energy – allowing millions of hectares to be returned to nature or to fibre, forage, or energy production. It could be done in ways that adequately reward farmers for feeding the world. But so far the political will to accomplish those results has been mainly lacking. The present reality is that in many parts of the world the soil, land, and nutrient sources of food are declining, and so are agricultural economies and communities" (Meadows et al. 2004, p. 66).

We started with reality, and then we presented utopian views and dreams. We will finish this part with a return to reality. Obviously there is great need for a new politics on the basis of Food Ethics in a globalized world – on the individual level and with respect for individual ethics in everyday life. A last remark reiterates some points concerning the "agrarian romanticism" as analyzed by the sociologist Klaus Eder (born 1946) in "The New Politics of Class. Social Movements and Cultural Dynamics in Advanced Societies" (Eder 1993). Eder writes from a historical point of view about the classical movements since Rousseau in western societies: "Regarding their effect on everyday life, two types of cultural movements were the most important ones: the communal and the vegetarian movements. The nineteenth-century experiments in country communal living were created to foster a closer relationship with nature. They mobilized people into leaving the cities. The concern with health dominated their relation to nature and practising horticulture gave them 'healthy' non-animal food. The movements for animal rights and the vegetarian movements promoted a genuinely social relationship with nature. The dialogue with nature that the romantics were dreaming of was practised and put into practice in these movements" (Eder 1993, p. 128). What can we learn from this analysis today? Is "romanticism" no more than anachronism? Eder's answer: "The discussion of the Romantic Movement and its derivatives shows that nature has since the beginning of modernity been the latent field of social and cultural struggles. Two factors have changed this latency. The ecological crisis has made nature the arena of public disputes. And, an increasing reflexivity in dealing with cultural traditions has led us to posit competing notions of relating to nature. Both have made nature a manifest and increasingly central field of social struggles in modern society. There are reasons to see it as becoming the field of a new type of class struggle replacing the old one, which only focused on the just distribution of goods in society. Such an emerging new class struggle would centre on another idea: on the idea of a more 'pure' nature, of an unpolluted environment" (Eder 1993, p. 129). Whatever this means today, two decades later the challenges concerning food and agriculture are far bigger than at any time before. We have to focus on a lot of problems and solutions in more detail, and the authors of this book try to do so. Nevertheless one main principle seems to be the same as in many kinds of former approaches to utopia: justice and fair trade. Concerning the fundamental relation between justice and autonomy today we need more theoretical and practical inspiration. The next part introduces a modern philosophical approach operating with "reason" and "reasons".

1.3 Reason & Justice as Common Principle for Reality and Utopia

Felix Ekardt (2004) makes a remarkable contribution to a universal and rational theory of justice. Closely linked, but at the same time different from several discourse-ethical trials by the philosophers Apel and Habermas, as well as from the fairness model of John Rawls, Ekardt proposes a rational background theory of justice. He is asking for the possibility of a coercive connection between reason, dignity, impartiality and freedom, which is universal and at the same time able to fill up these principles with well-defined content. Ekardt's first intention is to develop a liberal, discourse-rational theory of justice, which has universal validity and is at the same time intangible for critics of liberalism and of discourse-ethics. The starting point of his deliberations is a transcendental (in the meaning of Kant) fundament of reason, which is incontestable, as Ekardt shows in a conclusive way: Because – with regard to contents – there is no criterion for fairness and because of the de facto existence of heterogeneous opinions for what is just, there is not much else to do than to struggle with arguments about what might be possible. This process does not have a predetermined outcome. In the quest for justice, an open mind allows for reasons being put forward in an argumentative way. Thus, only this basic order can obviously be just, which facilitates the exchange of arguments. The arguments have to be presented in an orderly fashion, in which every participant is respected as free and equal, because only this basis allows for the possibility of open discussions. Ergo the principle of respect underlies all discourse. Ekardt's second argument for this fact has a transcendental structure: Whoever argues is trying to rationalize by resorting to "reason" by constructing sentences containing words like "because", "since", "for that reason", "therefore" and so on. A reasoned argument is thus convincing and can be reconstructed by everyone. A person who argues in a discussion, but then fails to pay due respect to other opinions, would act in a contradictory fashion, because this behaviour would be diametrically opposed to any professed logical argument. Arguments serve as a way of convincing others in an autonomous (in terms of self-legislating) way, and therefore this way of reasoning cannot be understood as a hierarchical act of someone getting into a conflict of opinion while arguing and assuming rationality, but only when denying the respondent the respect due to an autonomous being. Therefore, it is paramount to pay full respect to one's interlocutor when engaging in a rational argument. It makes no difference if the person is conscious of any reasonable implications or if the discussion is conducted merely as a means of receiving assurance, because the strong logical implications of our rationalizing would be at stake.

But because the participants are individuals and not collectives who operate rationally, the category "reason" insures respect for these autonomous individuals. Therefore the respect has to be valid for persons, not for collectives. Even more: the respect has to be valid for individual autonomy, because each individual has to be satisfied with the reasons put forth. And once justice is the topic of a discussion, both discourse and conclusion have to be generally acceptable and therefore impartial. This is the justification for the basics of liberty – and it is universal, because it ties in with human experience of rational discussion and therefore transcends all cultural borders.

And it is not only my present interlocutor who deserves respect and impartiality. If a reason-based discussion is initiated at all, the principles of respect and impartiality together with any inferred rights of freedom are automatically invoked also for any future discussants. Reasons are addressed to anyone who might be able to refute them. As soon as someone starts a rational discourse, anyone is invited to join in, and thus deserves respect. The central principles of discourse, respect and impartiality have to serve the dual purpose of being principles of action. On the basis of the fallibility of human decisions, respectively the openness of reason, discourse on a certain question might resume at any time. Anyone infringing upon the principles of dignity, respect, impartiality and autonomy beyond this discourse, would automatically place constraints upon any further discussions. Therefore these liberal principles must likewise apply to actions. On the basis of this, it can be said that principles per se are not part of rational thinking, thus creating the need for reasoned arguments proving what is equitable or inequitable. The reasons that are given may be put to the test at any time. In principle this allows new discourses to be opened at any time. However, this possibility would become obsolete, if respect and impartiality as universal basic elements of just societies would not apply to those actions occurring between discourses. Consequently respect and impartiality must be valid for procedural rules as well as for results of discourses. These are the core ideas of Ekardt's approach (Ekardt 2004), and the practical question is: how can we use these arguments for Food Ethics in a globalized world? We have to find "rational" solutions through forced rational dialogues about new solutions. In the past years, the so-called Consensus Conferences (Skorupinsky et al. 2006), i.e. small group sessions made up of laymen and experts have been one application of democracy at work. Today there are many democratic procedures and "ethical tools" on national and international levels, which have allowed for more and better bridges between reality and utopia (Beekman et al. 2006). They are waiting to be used by us.

References

Bacon F (1938) New Atlantis. In: Moore Smith CG (ed). Cambridge University Press, Cambridge
Baranzke H, Gottwald F-T, Ingensiep HW (2000) Leben Töten Essen. Anthropologische Dimensionen. Hirzel, Stuttgart

Beauchamp TL, Childress JF (2008) Principles of biomedical ethics (6th ed). Oxford University Press, Oxford

Beekman V et al. (ed) (2006) Ethical bio-technology assessment tools for agriculture and food production. Final report Ethical Bio-TA Tools (QLG6-CT-2002-02594). The Hague 2006 http://www.ethicaltools.info/content/ET1%20Final%20Report%20%28Binnenwerk%2059p%29.pdf. Accessed 8 January 2010

Butler S (1979) Erewhon. Erewhon revisited. Reprint (1872/1901). Dent, London

Campanella T (2004) The city of the sun. A poetical dialogue between a grandmaster of the knights hospitallers and a Genoese sea-captain, his guest. http://etext.library.adelaide.edu.au/c/campanella/tommaso/c18c/. Accessed 10 November 2008

Comstock GL (ed) (2002) Life science ethics. Iowa State Press, Ames, IA

Eder K (1993) The new politics of class. Social movements and cultural dynamics in advanced societies. Sage, London

Ekardt F (2004) Zukunft in Freiheit: Eine Theorie der Gerechtigkeit, der Grundrechte und der politischen Steuerung. CH Beck, München

Hesiod (1914) Works and days. (transl: Evelyn-White HG) http://www.sacred-texts.com/cla/hesiod/works.htm. Accessed 12 October 2008

Kropotkin P (1970) The conquest of bread. University Press, New York

Kropotkin P (1902) Mutual aid: A factor of evolution. William Heinemann, London

Meadows D, Randers J, Meadows D (2004) Limits to growth. The 30-year update. Earthscan, London

Meadows DS, Meadows DH, Zahn E (1972) Die Grenzen des Wachstums. Deutsche Verlags-Anstalt, Stuttgart

Mepham B (ed) (1996) Food ethics. Routledge, London

Mepham B (2005) Bioethics. An introduction for the biosciences. Oxford University Press, Oxford

Rousseau J-J (2004) A discourse upon the origin and the foundation of the inequality among mankind. http://www.gutenberg.net. Accessed 11 November 2008

Sachs W (1992) The developmental dictionary. A guide to knowledge as power. Zed Books, London

Schumacher EF (1999) Small is beautiful. Economics as if people mattered. 25 years later... with commentaries. Hartley & Marks, Vancouver

Schumacher EF (1973) Small is beautiful. Economics as if people mattered. HarperCollins, London

Skorupinski B, Baranzke H, Ingensiep HW, Meinhardt M (2006) Consensus conferences – A case study: PubliForum in Switzerland with special respect to the role of lay persons and ethics. J Agric Environ Ethics 20:37–52. doi 10.1007/s10806-006-9016-7

Part I
Food, Consumers, and Policy

Chapter 2
The Ethical Matrix as a Tool in Policy Interventions: The Obesity Crisis

Ben Mepham

Abstract The chapter addresses the issue of how governments of democratic states might seek to assist their citizens to reverse the serious trend towards overweight and obesity. Recent reports have stressed the contributory role of the obesogenic environment that characterises contemporary UK society, and suggested that to be effective policy interventions need to be implemented on several fronts. The chapter explores the multidimensional capabilities of the ethical matrix as both a procedural and a substantive tool in such programmes, focusing on food production, marketing and consumption.

2.1 Introduction

It is apparent that our rapidly changing world is now revealing, with startling regularity, a succession of developments which, although when viewed in retrospect seem to have been emerging gradually over a substantial period of time, now threaten to assume critical status within a few years. A prominent example is the impending epidemic of obesity, which in the UK is predicted to present a challenge comparable to that of global warming. Employing this as a case study, equally applicable to other advanced Western states, how should the UK government address this crisis? Recognising the seriousness of projected developments, but also conscious of the rights of individuals to choose their own lifestyles, in what ways and to what extent is it acceptable for the activities of citizens to be shaped by government policy?

As noted by Rawls (1993), modern liberal democracies are characterised by their accommodation of a plurality of reasonable, but to a degree, incompatible doctrines. While democratically elected governments may presume legitimacy for their

B. Mepham (✉)
Centre for Applied Bioethics, University of Nottingham, Nottingham, UK
e-mail: ben.mepham@nottingham.ac.uk

policy initiatives, it remains important, from both utilitarian and deontological perspectives, to ensure that interventionist policies affecting society as a whole are compatible with commonly accepted ethical standards. This is even more necessary where, as is often the case, governments are elected by a minority of the electorate. My aim in this chapter is to explore the value of a conceptual tool, the ethical matrix (EM), in addressing these issues.

The EM was introduced in 1996 in order to facilitate ethical deliberation and decision-making. On the basis of the notion of the common morality, its principal aim is to assist non-philosophers to appreciate the value of ethical insights in arriving at well-considered ethical judgements. In accord with the approach adopted by Beauchamp and Childress (2001) in the field of biomedical ethics, it appeals to prima facie principles, which are derived from both consequentialist and deontological theory. According to this approach, Fig. 2.1 illustrates a generic EM, which has relevance to decision-making in relation to food and agriculture: for use in particular circumstances the principles need to be "specified" according to the overall context. Accounts of the associated theory and practice are to be found in a number of publications, notably those by Mepham (1996, 2000a,b, 2001, 2005a, b, c, 2008), Mepham and Tomkins (2003) and Mepham et al. (2006). Other important applications include those of the Food Ethics Council (see http://www.foodethicscouncil.org/), Kaiser and Forsberg (2001), Chadwick et al. (2003); and critical appraisals of the method are provided by Schroeder and Palmer (2003) and Forsberg (2007). A response to some of these criticisms was published in Mepham (2004).

Respect for	Wellbeing	Autonomy	Fairness
Producers	Satisfactory income and work	Managerial freedom	Fair trade laws
Consumers	Safety and acceptability	Choice	Affordability
Treated Organisms	Welfare	Behavioural freedom	Intrinsic value
Biota	Conservation	Biodiversity	Sustainability

Fig. 2.1 A generic ethical matrix for use in issues concerning food and agriculture. Cell contents are specifications of the prima facie principles for each interest group

Since its inception, I have used the EM extensively; and it has also been used widely in collaboration with academic colleagues and by other groups. Not only has my own employment of the EM evolved and taken on different forms in different circumstances, but it has also been subjected to criticism and subsequent modification by other users. The result is that it is questionable whether use of the EM amounts to a methodology per se, or whether it is essentially an approach to ethical deliberation. But while no proprietary rights are claimed over the EM, and informed criticism is welcomed, it seems important to correct misapprehensions as to its intended use and value when these arise (see Mepham 2004), and when appropriate to provide stronger evidence to support its main tenets and develop its potential uses. The particular development here is concerned with ethical justification of certain public policy decisions. This is not an entirely novel type of application, because while in earlier exercises the focus was on ethical evaluations of new biotechnologies, rather than on their political implementation, some analyses have been more explicitly directed to policy issues. Even so, this chapter will explore a previously unexamined approach.

My earlier accounts of the EM have acknowledged the role in its development of ideas advanced by John Rawls – an approach which has, however, been challenged by Forsberg (2007). Here, I want to underline the perceived relevance of Rawls' thinking by examining the close association between my current ideas on the EM and the revised notion of justice as fairness, which Rawls expressed in a restatement of his theory, written shortly before his death. He began by identifying four roles or objectives of political philosophy, which may be summarised as follows:

(1) provision of "a focus on deeply disputed questions, to see whether, despite appearances, some underlying basis of philosophical and moral agreement can be found"
(2) provision of "a unified framework within which proposed answers to divisive questions may be made consistent and the insights gained from different kinds of cases can be brought to bear on one another ..."
(3) recognising the "fact of profound and irreconcilable differences in citizens' reasonable comprehensive and philosophical conceptions of the world, and their views of the moral and aesthetic values to be sought in human life" to try to reconcile them "by showing the reason and indeed the political good and benefits of (such reconciliation)"
(4) promotion of the view that "political philosophy is realistically utopian; that is, (it probes) the limits of practicable political possibility". (Rawls 2002)

From my perspective, the structure, aims and modes of use of the EM may reasonably be said to resonate with these roles and objectives, as defined by Rawls. Thus, referring to the above numbering, in several diverse settings the framework (2) has proved valuable in focusing on contentious issues (1), by employing a strategy which assesses how far ethical ideals are met by proposed changes (4) and sometimes discovering an "overlapping consensus" (Rawls 2002) despite marked differences in people's moral values (3). According to this view, the starting point for ethical analysis is the formulation of a set of prima facie principles that are deemed evident in the common morality, but for which no weighting is assigned, and consequently to which prospective users can attach as much significance as is deemed appropriate (which in some cases might be "zero"). According to Gillon (1998), the principles used in the EM provide "a transcultural, transnational, transreligious, transphilosophical framework for ethical analysis" by allowing differences of emphasis within a scheme of universal applicability. Despite the suggestion that structuring the process of deliberation by employing these principles may bias outcomes (e.g. Fraser 2001), no compelling alternative approaches appear to have been proposed. On the contrary, most feedback from participants in workshop exercises has been strongly positive (Mepham and Millar 2001).

The EM has been used by individuals and groups in numerous settings. Forsberg (2007) appears to suggest that the only valid way of using it is as a tool employed in a deliberative process involving a wide range of stakeholder representatives (designated the "bottom-up" approach in Mepham et al. 2006). Such exercises are certainly valuable. Indeed, the practice was initiated with colleagues at Nottingham (e.g. Mepham and Millar 2001); but they do not circumscribe the usefulness of the EM. Moreover, they encounter many practical constraints, such as those relating to time, cost and selecting appropriately representative groups of participants. In the current context, the aim is to explore use of the EM by policy-makers in formulating public policy. Of necessity, the decisions they make need to take account of a range of political and economic considerations that might well exceed the competence of many stakeholders to assess adequately.

2.2 Obesity in the UK

As noted by a UK government Foresight report (Department for Innovation, Universities and Skills 2007) "Being overweight has become a normal condition, and Britain is now becoming an obese society". It does not appear to be a question of people having less willpower than earlier generations, or having become more gluttonous, but rather of profound changes having occurred in society over the last 50 years – which have impacted on work patterns, transport, food production and sales, recreational and leisure activities. Such changes have exposed the underlying biological tendency for many people to put on weight and to retain it, so that currently 25% of adults are assessed, according to their body-mass index (BMI) as "obese".[1] And while that might seem to be merely a cosmetic problem, in fact, overweight and obesity increase the risk of a wide range of chronic disease conditions, such as hypertension, cardiovascular disease (including stroke), type 2 diabetes and cancer. Wellbeing can also be seriously impaired by physical incapacity, social stigma, low personal esteem and a generally low quality of life. Such trends are evident in many countries, and are already perhaps most pronounced in the USA (where 39% of the population is classed as obese), but within the EU the UK is at the forefront of this regrettable trend, with authoritative predictions suggesting that 60% of Britons will be obese by 2050.

Not only are there serious consequences for individuals affected by obesity, but the economic implications are also projected to be substantial. For example, in the UK the costs to the National Health Service (NHS) attributable to overweight and obesity are predicted to double to £10 billion p.a. by 2050. But these are not the

---

[1] Obesity is generally assessed on the basis of the *body mass index* (BMI), defined as body weight (kg) divided by height (m).[2] The World Health Organisation (WHO) defines "normal weight" as a BMI of 18.5–24.9, "overweight" as a BMI of 25.0–29.9, and "obese" as a BMI of over 30.0.

major costs, because the wider financial impacts on society and businesses (e.g. in terms of lost productivity) are anticipated to reach (at current prices) £50 billion p.a. (Department for Innovation, Universities and Skills 2007).

Clearly, at the physiological level, obesity is a consequence of more food energy being consumed than is expended in physical activity. This excess energy is laid down as fat, and when the process continues over a substantial period the result is an increase in BMI, which is first manifest as overweight and then as obesity. When, as in the immediately post-war period of the 1950s, there was a shortage of food – and physical energy was expended in activities such as manual work, cycling and outdoor sports, the way in which most people lived was conducive to weight maintenance within the "normal range". By contrast, in the UK most people now live in an obesogenic environment, in which work and leisure activities are largely sedentary, much transport is motorised (while escalators and lifts have replaced stairs), and low food prices and persuasive advertising (together with peer group pressure) encourage overeating of high sugar, salt and fat foods (hereafter referred to as HSSFF). It is evident that the problem of obesity is highly complex, and not likely to be effectively addressed by simply exhorting people to "eat wisely". Rather, the fact that the drivers of the condition are deeply embedded in the way modern society has been constructed would seem to call for comprehensive approaches entailing government intervention. The recognition of this social responsibility has led recently to a number of prominent UK enquiries (e.g. Department for Innovation, Universities and Skills 2007, Nuffield Council on Bioethics 2007,[2] Sustainable Development Commission 2008, Department of Health 2008).

2.3 An Ethical Approach to Policy Decisions

Of the above reports, that of the Nuffield Council on Bioethics is the only one to specifically address obesity (and other public health issues) from an ethical perspective. Its insights provide some valuable guides to how policy might be shaped, which it is an aim of this chapter to augment, amplify and, where appropriate, criticise. The justification for this objective, when so many reports on obesity have already been published, is that of providing specifically ethical grounds for policy decisions that appeal to principles encompassed by the common morality.

The Nuffield report emphasises the role of a stewardship model according to which governments have an obligation to provide conditions that allow their citizens to be healthy – part of which entails efforts to reduce inequalities in health within the population. While recognising the need to respect personal choice (e.g. by avoiding coercive or intrusive measures that have not received appropriate consent), the Nuffield report considers that in order to protect the vulnerable (e.g. children and disadvantaged adults) various policy interventions may be ethically justifiable.

---

[2] The author's submission to the Nuffield Council on Bioethics public consultation exercise may be viewed at: http://www.nuffieldbioethics.org/fileLibrary/pdf/Professor_Ben_Mepham001.pdf

Depending on the seriousness of the issue, the authors of the report consider that the options implemented might be best considered as "rungs on an intervention ladder", with the least intrusive entailing no more than monitoring a situation (e.g. the incidence of obesity in different socioeconomic groups) while the most intrusive might entail legal measures to ban a certain foodstuff.

These are useful perspectives, but they omit some important considerations. Use of the EM can clarify the situation by identifying what ethical concerns are at stake at different points in the food chain, from "field to fork", and how the needs of different stakeholders would be affected by specific policy interventions. My previous use of the EM has concentrated on its heuristic potential, by means of which users are invited to specify the principles in ways appropriate to their interpretation of the issues and assess anticipated impacts of innovations on the degree of respect these principles are accorded. Readers of this chapter are referred to the references listed above for fuller accounts of the theory and recommended practice in using the EM. Others have amended the approach by employing a project matrix (Chadwick et al. 2003) or distinguishing between a value matrix and a consequence matrix (Forsberg 2007) – changes which can lead to substantially different approaches, for example in which the cells instead of specifying principles contain questions (e.g. Forsberg 2004).

Here, I introduce two new forms of matrix, each applicable at a different level of an ethical analysis. These are:

- a specified principles matrix (SPM): see Fig. 2.2, and
- a policy objectives matrix (POM): see Fig. 2.3

Respect for	Wellbeing	Autonomy	Fairness
Producers of food: farmers and associated workers; food manufacturers; food processors	Satisfactory income and work from producing less obesogenic food	Self-determination	Fair trade laws
Marketers of food: wholesalers; retailers; restauranteurs; advertisers	Satisfactory income and work from selling less obesogenic food	Free market	Fair trade laws
Consumers at risk of obesity	Reduced risks of obesity and associated diseases	Informed food choice	Equality of opportunity e.g. in access to healthy food
Society members	Health and prosperity of global population	Diversity	Sustainability

Fig. 2.2 A proposed specified principles matrix (SPM) pertinent to addressing obesity, which specifies the ideals which policy decisions concerning food production, marketing and consumption might aspire to respect

Respect for	Wellbeing	Autonomy	Fairness
Producers of food: farmers and associated workers; food manufacturers; food processors	Legislate/regulate to significantly reduce production of obesogenic foods, by diverting production to healthier products	Protect innovative and entrepreneurial practices	Promote fair trade
Marketers of food: wholesalers; retailers; restauranteurs; advertisers	Legislate/regulate to significantly reduce sales of obesogenic foods, by diverting sales to healthier products	Protect innovative and entrepreneurial practices	Promote fair trade
Consumers at risk of obesity	Promote healthier lifestyles to avoid obesity, by encouraging healthy eating	Cultivate informed food choices through education	Ensure equality of access to healthy food and nutrition education, so promoting a healthier lifestyle
Society members	Provide facilities to promote wellbeing	Promote toleration of infirmity and disability through education and non-discriminatory practices	Ensure availability of sustainable supplies of healthy food through long-term planning

Fig. 2.3 A policy objectives matrix (POM) indicating some (speculative) proposals relating to food production, marketing and consumption

With reference to these figures, the abbreviations used below identify specific cells (e.g. SPM/PW refers to producer wellbeing in the specified principles matrix).

In the SPM the object is to identify (groupings of) coherent interest groups and the idealistic objectives to which observing the principles (respect for wellbeing, autonomy and fairness) might reasonably aspire – in line with the Rawlsian strategy identified above (no. 4). But since these are prima facie principles it is almost inevitable that some will take precedence over others, a situation which typically characterises policy decisions, and demands transparent justification. The compositions of the different groups are based on the concept of broadly similar objectives, for example in the case of producers, all are involved in the growing and manufacture of food, including associated activities, such as agrochemical production, butchery and haulage: some people in these categories will live in less developed countries.

Clearly, there are wide differences in the circumstances of different groups identified, to which policy makers would need to pay due attention. In the form of EM employed by Kaiser and Forsberg (2001), selected participants in a deliberative exercise were invited to construct a matrix by deciding how respect for the different principles (which were, however, proposed by the organisers) was to be specified. The risk in such a procedure is that participants may limit their perceptions of desirable outcomes to what seems "realistic" in the prevailing circumstances.

In contrast, reference to idealistic principles has more chance of guaranteeing that ethical criteria are prioritised, rather than marginalised by overriding practical limitations. In line with Rawlsian principles (see no. 2 above) the explicit statement of specified ethical principles facilitates the possibility of discovering whether "some underlying basis of philosophical and moral agreement can be found". Of course, ultimately, practical considerations inevitably play a major role in policy-making, but according to this analysis their consideration is appropriately deferred until a later stage.

Even so, it must be conceded that appealing to "ideals" allows room for significant differences of interpretation. For example, the term "satisfactory" with reference to income (Fig. 2.2, SPM/PW and SPM/MW) begs the question of how this criterion should be defined. It is clearly impossible to stipulate absolute standards, so that perceived comparability with other incomes is likely to influence notions of "satisfaction". Another consideration is the extent to which people engaged in activities now considered harmful (such as producing or selling HSSFF) are culpable of irresponsible behaviour, or whether the market arrangements that have allowed them to lawfully engage in such activities amount to ethical endorsement. In the latter case, governments might be said to have a duty of reparation if policy changes were to penalise them.

Society members is a term employed to acknowledge the fact that obesity is a condition only affecting a proportion, albeit an increasingly significant proportion, of (the global) society, but that respect for wellbeing, autonomy and fairness demands equal attention for everyone. In the current formulation (which it is emphasised is intended to be illustrative rather than definitive), the interests of future generations of living beings are included (hence the significance of sustainability in cell SF). It might also be considered appropriate for this interest group to include ethically considerable non-human living organisms (which were designated "biota" in Fig. 2.1). However, their marked differences from humans might reasonably suggest to most people that they should be assigned to a separate category – and they are excluded from Fig. 2.2.

In summary, the SPM proposes a way in which fundamental ethical principles, which were presented differently in Fig. 2.1, may be specified in the context of policy issues relating to obesity. In that the specifications are highly generalised it might be anticipated that they would find support from a substantial majority of people, even acknowledging the "plurality of reasonable, but to a degree, incompatible doctrines" that characterises modern liberal democracies.

2.4 Impacts on Policy Formulation

A central plank of the Nuffield report's analysis was the need to protect vulnerable groups (e.g. infants, senile people, those suffering from addictions) from harm, a motive that is represented in one form of POM (Fig. 2.3) by cells CW, CA and CF. The primary aim here is to propose a structured framework by which ethical principles may be translated into policy objectives. Representation in a matrix facilitates assessment of the relative ethical claims of competing interests, and can make explicit the weighting that is deemed appropriate in the subsequent formulation of policies. By bringing ethical considerations to the fore, it acts as a substantive ethical tool; and by requiring policy-makers to articulate their assessments of impacts on each cell of the matrix, it acts as a procedural tool.

It is important to note that Fig. 2.3, which is limited to factors directly affecting food production, marketing and consumption, represents just one of a number of forms of POM that would need to be constructed to comprehensively address different aspects of the obesity crisis. For example, the obesogenic environment is susceptible to amelioration by numerous policies, which relate, inter alia, to provision of sports facilities, cycle tracks, modification of the design of buildings, and educational curricula at both school and adult levels. Moreover, when the ethical principles are universalised, so as to apply to the global society (perhaps including morally considerable non-human living organisms), it becomes apparent that measures that seek to address obesity should be considered alongside all other considerations affecting life on Earth, now and in future.

The POM illustrated in Fig. 2.3 seeks to define relevant ethical principles in policy-oriented terms that address issues concerning food. But the specifications clearly fall far short of stipulating any actual policy recommendations. This follows from the observation of Gillon (1998) that the principles "are general guides that leave considerable room for judgement in specific cases and that provide substantive guidance for the development of more detailed rules and policies". Thus, it might appear that the analysis has little direct impact on the process of ethical policy formulation. However, that assessment would overlook several important considerations.

Firstly, such an analysis should provide invaluable stepping stones to the achievement of a Rawlsian "overlapping consensus" (Rawls 2002). In contrast, very few policy decisions to date have attempted to appeal to ethical principles, other than (probably unwittingly) to a form of utilitarian cost/benefit analysis. Secondly, whether or not consensus proves possible, appropriate use of the SPM and POM can provide explicit ethical justification for whatever policy decisions are ultimately made. This is because, when used conscientiously, the EM (in the various forms described here) is capable of facilitating the formulation of judgements that are comprehensive in scope, explicit in articulation, transparent in terms of their justification, and arrived at by a process of rational deliberation. But, thirdly, because ethical considerations necessarily permeate all subsequent policy decisions (although this is not widely appreciated), subsequent use of forms of the EM can have an explicit bearing on how the specified policy proposals in Fig. 2.3 might be implemented. This claim will now be examined.

2.5 Deciding on Ethical Policies

The Nuffield report refers to a "ladder of policy intervention", whereby governments in the interests of acting as stewards of their citizens' interests may legitimately take increasingly intrusive measures (e.g. affecting food producers, on the one hand, or obese individuals, on the other) to rectify adverse impacts on public health. What this implies, although it is not stated explicitly, is that political change is to be effected by the exercise of power. In his illuminating work "The Anatomy of Power", JK Galbraith (1984) identified three ways in which power is exerted (e.g. by individuals, companies and governments):

- condign power: the imposition of undesirable consequences if behaviour is not changed as required
- compensatory power: rewarding (often financially) those who accede to requests to change behaviour
- conditioned power: the exercise of persuasion and education to change behaviour

Each has its malign aspect (represented e.g. as malicious threats, bribery and brainwashing, respectively), but each also can (and perhaps, must, at some level) play a benign role in policy formulation. However, from an ethical perspective, it is important to consider whether recommendations to exert political power by any of these means are justifiable in terms of their impacts on respect for principles specified in the EM.

For example, with reference to Fig. 2.3, the interests of children assume paramount importance because of their intrinsic naivety and hence vulnerability to advertising campaigns that promote consumption of HSSFF, coupled with the high risk that early onset obesity will become a permanent condition. In such a case, respect for advertisers' prima facie rights to exercise "innovative and entrepreneurial practices" (Fig. 2.3, cells PA and MA) might well be considered justifiably overridden by necessary measures taken to respect CW, CA and CF. It seems that achieving the desirable outcome of reducing childhood obesity rates will inevitably entail the exercise of (or some combination of) condign, compensatory or conditioned power. The question is: "Can this be achieved while also respecting PF and FF (Fig. 2.3)"?

Examples of strategies aimed at reducing childhood obesity are: (i) requiring schools to introduce low HSSFF menus, (ii) banning sale of HSSFF in the vicinity of schools and in school vending machines, (iii) restricting advertisements of HSSFF on television, and (iv) subsidising retailers to sell fresh fruit and vegetables. These are all what have been called "command and control" regulations, which imply that professional health experts are certain of the answers to the obesity problem. Given the failure of such strategies in the past, the effectiveness of this approach is highly questionable. However, a sounder policy, from both ethical and practical perspectives, may be performance-based regulation, which assigns significant responsibility for causing obesity to the large food companies that sell HSSFF (Sugarman and Sandman 2007). According to this approach, companies would be required to "put their own house in order" by reducing HSSFF in proportion to the extent to which they are calculated to have contributed to the problem. Internalising cost in this way, by requiring manufacturers to bear the cost of their activities (in a manner analogous to the "polluter-pays" principle) is a practice used by economists to justify industry regulation; and it has the additional benefit of reducing the burden on taxpayers in a way that could be seen to respect the principle of fairness. It also has the advantage that industry's "innovative and entrepreneurial skills" (Fig. 2.3) will be given full reign within a commonly agreed constraint (i.e. on a "level playing field"). Various forms of POM might here be valuable tools in arriving at ethically sound policies.

Another instance, which was overlooked in the Nuffield report, where policy decisions might make a major impact on public health, concerns the production of HSSFF. Partly, this is a result of agricultural practices that are financially rewarded for producing fatty foods, mostly of animal origin (Crawford and Ghebremeskel 1996), and partly it is due to food manufacture and processing practices. According to a recent report, in the USA the supply of food exceeds the need for it by about 50% (Tong 2004), and while no comparable data have been obtained for the UK it seems unlikely that the situation is substantially different. To avoid concentrating on "end of pipe" solutions, measures to limit production of excess, and excessively unhealthy, food supplies, would seem to be an important aspect of policies aimed at countering obesity. Here again, structuring analyses on forms of POM may prove valuable.

Finally, it is important to appreciate that the definition of the boundaries of any ethical analysis exerts an overriding influence. For example, in Fig. 2.3 the somewhat opaque expression Society members is capable, as indicated above, of being taken to represent the global society as a whole (including human and non-human lives, now and in future) or, as is perhaps more usually the case, members of a nation state – which it might be considered presents a task that is challenging enough on its own. Yet increasingly we are being forced to address problems from a global perspective, notably in dealing with the threats posed by global warming. So it would not be surprising to discover, in view of the increasing globalisation of food markets, that the only authentic approach to devising an ethical food policy, is one which recognises that the millions of people suffering malnutrition as a result of obesity are matched by an equal number of malnourished people who are chronically underfed. Consequently, respect for global justice would seem to demand nothing less than a global ethical food policy (see Follesdal and Pogge 2005).

2.6 Conclusions

In this chapter I have argued that to win support for policy interventions aimed at reducing obesity, they need to be formulated on the basis of ethical decisions that prioritise public health. In conformity with recommendations of the Nuffield report, the UK government should adopt a stewardship model to protect the interests of the most vulnerable and seek to achieve equality in society, without overruling responsible personal choice. Use of the EM in the forms outlined here (as specified principles- and policy objectives-matrices) could provide a conceptual tool (serving both substantive and procedural roles) to arrive at transparent and ethically justified public policy decisions.

The chapter has both bold ambitions and modest ambitions. They are bold in suggesting that all policy decisions that involve consideration of social, biological and environmental concerns should entail a form of ethical evaluation that would be facilitated by employing appropriate ethical tools. They are modest because, within the limitations of a short article, the proposals concerning the use of the EM as such a tool have been merely suggestive of a mode of analysis, leaving many questions unanswered.

References

Beauchamp TL, Childress JF (2001) Principles of biomedical ethics (5th edn). Oxford University Press, New York and Oxford

Chadwick R, Henson S, Moseley B, Koenen G, Liakopoulos M, Midden C, Palou A, Rechkemmer G, Schröder D, von Wright A (2003) Functional foods. Springer, Berlin

Crawford M, Ghebremeskel K (1996) The equation between food production, nutrition and health. In: Mepham, B (ed) Food ethics. Routledge, London, pp. 64–83

Department for Innovation Universities and Skills (2007) Foresight, tackling obesities: Future choices-project report. Department for Innovation, Universities and Skills, London

Department of Health (2008) Healthy weight, healthy lives: A cross-governmental strategy for England. Department of Health, London

Food Ethics Council (2008) http://www.foodethicscouncil.org

Follesdal A, Pogge T (eds) (2005) Real world justice. Springer, Dordrecht

Forsberg E-M (2004) Ethical assessment of marketing GM roundup ready rape seed GT73. In: de Tavernier J, Aerts S (eds) Science, ethics and society, preprints of 5th EURSAFE congress. EURSAFE, Leuven, Belgium, pp. 177–180

Forsberg E-M (2007) A deliberative ethical matrix method – Justification of moral advice on genetic engineering in food production [Dr Art dissertation], University of Oslo

Fraser (2001) What is the moral of the GM food story? J Agric Environ Ethics 14:147–159

Galbraith, JK (1984) The anatomy of power. Corgi, London

Gillon R (1998) Bioethics: Overview. In: Chadwick R (ed) Encyclopedia of applied ethics, vol. 1. Academic Press, San Diego, pp. 305–317

Kaiser M, Forsberg E-M (2001) Assessing fisheries – Using an ethical matrix in a participatory process. J Agric Environ Ethics 14:191–200

Mepham B (1996) Ethical analysis of food biotechnologies: An evaluative framework. In: Mepham B (ed) Food ethics. Routledge, London, pp. 101–119. Reproduced in: Chadwick R, Schroeder D (eds) (2002) Applied ethics: Critical concepts in philosophy. Routledge, London, pp. 343–359

Mepham B (2000a) The role of food ethics in food policy. Proc Nutr Soc 59:609–618

Mepham B (2000b) A framework for the ethical analysis of novel foods: The ethical matrix. J Agric Environ Ethics 12:165–176

Mepham B (2001) Novel foods. In: Chadwick RF (ed) The encyclopedia of ethics of new technologies. Academic Press, San Diego, pp. 299–313

Mepham B (2004) A decade of the ethical matrix: A response to criticisms. In: de Tavernier J, Aerts S (eds) Science, ethics and society, preprints of 5th EURSAFE congress. EURSAFE, Leuven, Belgium, pp. 271–274

Mepham B (2005a) The ethical matrix as a decision-making tool, with specific reference to animal sentience. In: Turner J, D'Silva J (eds) Animals, ethics and trade: The challenge of animal sentience. Earthscan, London, pp. 134–145

Mepham B (2005b) Food ethics. In: Gunning J, Holm, S (eds) Ethics, law and society, vol. 1. Ashgate Publishing, Aldershot, pp. 141–151

Mepham B (2005c) The ethical matrix: A framework for teaching ethics to bioscience students. In: Marie M, Edwards S, Gandini G, Reiss M, von Borell E (eds) Animal bioethics: Principles and teaching methods. Wageningen Academic Publishers, Wageningen, pp. 313–327

Mepham B (2008) Bioethics: An introduction for the biosciences (2nd edn). Oxford University Press, Oxford

Mepham B, Millar KM (2001) The ethical matrix in practice: Application to the case of bovine somatotrophin. In: Pasquali M (ed) Food safety, food quality, food ethics, preprints of 3rd EURSAFE congress. A&Q, University of Milan, Florence, pp. 317–319

Mepham B, Tomkins SM (2003) Ethics and animal farming (web exercise). Compassion in World Farming Trust, Petersfield. www.ethicalmatrix.net

Mepham B, Kaiser M, Thorstensen E, Tomkins S, Millar K (2006) Ethical matrix: Manual. http://www.ethical.tools.info/content/ET2%20Manual%20EM%20(Binnenwerk%2045p)pdf

Nuffield Council on Bioethics (2007) Public health: Ethical issues. Nuffield Council on Bioethics, London

Rawls J (1993) Political liberalism. Columbia University Press, New York and Chichester, Sussex

Rawls J (2002) Justice as fairness: A restatement. In: Kelly E (ed). Belknap Press, Cambridge, Mass and London, England, pp. 3–4

Schroeder D, Palmer C (2003) Technology assessment and the 'ethical matrix'. Poiesis Prax 1: 295–300

Sugarman SD, Sandman N (2007) Fighting childhood obesity through performance-based regulation of the food industry. Duke Law J 56:1403–1490

Sustainable Development Commission (2008) Green, healthy and fair. Sustainable Development Commission, London

Tong R (2004) Taking on 'big fat': The relative risks and benefits of the war against obesity. In: Boylan M (ed) Public health policy and ethics. Kluwer, Dordrecht

Chapter 3
Ethical Traceability for Improved Transparency in the Food Chain

Christian Coff

Abstract Some practices in the agri-food sector worry consumers. Consumers might for instance be concerned about animal welfare, health, environmental issues, transparency of the food chain and so forth.

A question, which confronts consumers today, is how they can become capable of acting upon such ethical concerns. Information is often seen as an answer to the mentioned consumer concerns. Paradoxically, although consumers are bombarded with information on food – from the media, the food industry, food authorities, NGOs and interest groups – details about how foods are actually produced is often hard to find. Much of the information available is superficial, conflicting or partial, and it is hard for consumers seeking to make informed food choices on ethical matters to know which information to trust. Food traceability, which provides a record of the history and journey of a given food, and which is increasingly used in the food sector for legal and commercial reasons, has the potential to communicate a more authentic picture of how food is produced and thus provide an answer to some consumer concerns.

3.1 Introduction

Some practices in the agri-food sector worry consumers. Consumers might for instance be concerned about animal welfare, health, environmental issues, transparency of the food chain and so forth.

A question, which confronts consumers today, is how they can become capable of acting upon such ethical concerns. Information is often seen as an answer to the mentioned consumer concerns. Paradoxically, although consumers are bombarded with information on food – from the media, the food industry, food authorities, non-governmental organisations (NGOs) and interest groups – details about how

C. Coff (✉)
University College Sjælland, Ankerhus, Denmark
e-mail: coff@c.dk

foods are actually produced is often hard to find. Much of the information available is superficial, conflicting or partial, and it is hard for consumers seeking to make informed food choices on ethical matters to know which information to trust.

Food traceability, which provides a record of the history and journey of a given food, and which is increasingly used in the food sector for legal and commercial reasons, has the potential to communicate a more authentic picture of how food is produced and thus provide an answer to some consumer concerns. The idea of "ethical traceability" adapts the idea of food traceability to record and communicate the ethical aspects of a food's production history. It offers a mechanism for communicating more comprehensively and reliably the information about food production practices that consumers need in order to be able to make food choices consistent with their values. Used imaginatively, it could also provide an opportunity for two-way communication along food chains, allowing the views of consumer-citizens to be taken into account along the length of the chain.

As food traceability retells the history of a food, it can address the ethical aspects of that history, enabling more informed food choice. Secondly, it can act as a (democratising) means for enabling food consumers to participate more fully as citizens in the shaping of the contemporary food supply chain.

3.2 The Use of Traceability in Contemporary Food Chains

Today, traceability has become a common practice in the agri-food sector. Indeed, since 2005 EU law has required a certain level of traceability on the part of all food operators in the EU. Table 3.1 maps the key applications and objectives of traceability in contemporary food systems.

The first four levels are widely used, while the fifth objective – consumer information and communication – is still in its initial phase. As we shall see, this fifth objective is, however, essential for developing traceability in the ethical direction of consumers' informed food choice. The first category, risk management and food safety, has been a primary focus of regulatory attempts to introduce traceability. Food safety control has been built upon process-based auditing, such as HACCP (Hazard Analysis of Critical Control Points) standards and the International Organisation for Standardisation ISO 9000 (ISO 2007). The need to be able to recall contaminated products for public health reasons motivated food producers to incorporate traceability systems into supply chain management processes originally implemented to achieve efficiencies (Farm Foundation 2004, p. 8). The latter – supply chain management and efficiency – is the third category in Table 3.1, and its main concern is to allow food companies, notably the corporate retailers, to manage the flow of goods and information, link inventory to consumer purchasing, set product specifications for growers and processors under contract, and so on, in order to meet market demand and secure the efficient use of resources. Traceability is thus an instrument that can be deployed for a variety of purposes, often at the same time. Hence in practice there will usually be some overlap between the different categories

Table 3.1 Key functions of traceability in the food sector (Coff et al. 2008b)

(i) Objectives of food traceability

1. *Risk management and food safety*
   - Risk assessment: mapping of foods and feed, food ingredients and processing technologies that have food safety implication (e.g. hygiene)
   - Food residue surveillance: food sampling at appropriate points testing for residues, e.g. pesticides
   - Public health recall systems: identification of breakdowns in food safety along the food supply chain, allowing recall of contaminated products for the purpose of protecting public health

2. *Control and verification*
   - Surveillance and auditing of producer and retailer activities
   - Avoidance of fraud and theft: control of products by chemical and molecular approaches (biological "food-prints")
   - Identification of responsible actors (but also claims of innocence!)
   - Ingredients definition
   - Avoidance of negative claims (e.g. "may contain GMO traces")

3. *Supply chain management and efficiency*
   - Cost effective management of the supply chain
   - Computerised stock inventory and ordering systems linked to point of sale
   - Just-in-time delivery systems
   - Efficient use of resources (cost minimisation)

4. *Provenance and quality assurance of products*
   - Marketing of health, ethical and other claims
   - Authenticity: identity of the product (food authentication) and the producer
   - Typicality: as with European schemes for Protected Designation of Origin (PDO) and Protected Geographical Indication (PGI)
   - Quality assurance of standards at different stages of production and/or processing (e.g. environmental protocols for production)
   - Final product quality assurance

5. *Information and communication with the consumer*
   - Transparency of the production history
   - Facilitation of informed food choice, through transparency and the ability to compare different products
   - Recognition of specific consumers concerns and information demands – where such concerns and demands are not static but may evolve
   - Public participation; consumer services, companies' "care lines" and consultation to obtain consumer feedback

depicted in Table 3.1, and traceability will rarely if ever be implemented for only one of the objectives mentioned. For example, the second category interweaves with the other two mentioned above. The record of the production history of a product can be used for surveillance and fraud prevention. It is interesting to observe that the two largest retail companies of the world, Wal-Mart and Carrefour, are increasingly asking for complete traceability from their suppliers (Bantham and Duval 2004).

The fourth category is likewise linked to the second, as it concerns verification of quality claims and label schemes. Quality is a complex term, as the perception and dimensions of quality are continually shifting. But the goals of traceability as set out in the fourth category can to some extent reflect ethical criteria for food production practices and also consumers' ethical concerns, and communicate them via labels. Four examples, out of many labelling schemes, are: the organic labels found in most European countries, the UK Red Tractor scheme, the UK Royal Society for the Protection of Animals' (RSPCA) Freedom Food label, and the French Label Rouge.

The fifth category of objectives for traceability is far from fully developed. In some ways it is an aspiration that would facilitate consumers' understanding of food production practices and their ability to make informed choices about the foodstuffs they purchase and eat. It concerns the *communication* of production practices in the food chain. In this sense, traceability is about visibility; it is about making the production history of food visible to the eyes of the consumers. It allows producers and retailers to establish a more advanced kind of communication with consumers about production practices.

To some extent, all traceability is ethical. Food safety is obviously an ethical issue since it aims at protecting consumers from food-borne diseases and pollution. Preventing fraud in the food chain is likewise inherently ethical, as is guaranteeing the accuracy of the information provided to consumers, and the verification of assurance and labelling schemes. However, it is at the communication level that specifically "ethical" traceability gains a certain power. For actors in the food chain (be they processors, manufacturers or retailers) who wish to secure a minimum level of ethical behaviour among their suppliers, ethical traceability provides information on the ethics of a given product's production history, which is essential if the buyer is to be able to form an ethical judgement of the supplier (Coff 2006). And the same goes for consumers: ethical traceability should provide the information necessary for consumers to exercise their ethical judgement about the production history of a given food, and thus allow consumers' informed choice.

3.3 Traceability and Food Ethics

Most people are aware that in recent decades massive changes have taken place in agricultural and food production practices. This is clear not only from the radical changes that can be observed in the landscape and the bewildering array of goods available in contemporary supermarkets, but also from media headlines. Media coverage of food production practices tends to highlight negative aspects, such as food scandals, environmental and animal welfare problems, and so forth.

The physical, social and mental separation of production and consumption, which is characteristic of modern societies, means that in most cases producers and consumers do not know each other and that consumers do not know what happens during production processes. They are invisible to one another. In spite of this differentiation of the two spheres, and the obscurity of the food system, people, as citizens and consumers, may still seek to feel that they somehow are involved in agriculture and food production. Or, at least, that food production practices matter, in the sense that it makes a difference to consumer-citizens if food is produced in one way as opposed to another. But how can food production practices matter, even though production has been so clearly separated from consumption?

There is an old saying that "if you eat, you are involved in agriculture". Shopping, preparing and eating are key notions for understanding the involvement of consumers in the agri-food sector. These three activities are acts that lead the thoughts in two different directions. We could say that our thoughts are led both backwards and forwards in time. Shopping, preparing and eating are, so to speak, specific points in a chain of events, from which it is possible to think both backwards and forwards. We think backwards when we consider *what* we are buying, preparing and eating. We cannot ascertain if something is edible, and for instance whether it is meat or a vegetable, without relating our sensuous perception to our knowledge of what is meat and what is a vegetable. We have an idea of what is meat and what is a vegetable, and from our experience we judge them to be either edible or non-edible. Now, this experience is often associated with many different stories. One very simple story is that meat comes from living animals. For some people, this knowledge is very important (vegetarians, for instance). This simple illustration shows how, in the act of eating, we consciously or unconsciously direct our thoughts towards the past.

We also direct our thoughts towards the future, when considering how a food will affect our bodies. Food is taken into the body. It is incorporated and incarnated. In this sense, food links our body with past events in the agri-food sector in a very physical sense. But we might also consider the pleasure of the food (taste, digestibility, effect on the mental state, and so forth), the healthiness of the food, and the social and cultural contexts of shopping, cooking and eating. With the food we choose, we make a statement about our identity and connect ourselves with other people who make the same kind of food choices. Vegetarianism, or whatever kind of diet we choose, is as much about belonging to certain groups as about eating. This gives some idea of why production practices in the agri-food sector still matter for many people, despite the separation of production and consumption.

To return to the meaning of the concept of traceability, it refers to the history of a product and the records kept of that history. The internationally most recognised definition of traceability belongs to ISO (ISO 2007). ISO 9000:2000 refers to a set of quality management standards. To this set belongs ISO 8402. This standard defines traceability as "the ability to trace the history, application or location of an entity by means of recorded identifications". Thus, traceability seems to offer the possibility of making the link between production and consumption visible. As mentioned earlier, traceability is already a requirement of EU food law (regulation (EC) No 178/2002). Thus, all European food businesses have a legal responsibility to implement traceability. The question raised here is: can this traceability be used to provide ethical information to consumers about the production history of foods, and thereby enable consumers to make informed choices on ethical issues?

A central concern, then, is how traceability can link to food ethics. Food ethics is a discipline that embraces many different ethical and philosophical studies on food. However, it is not just an academic discipline; it also describes more practical ways in which people think about food and act according to their values of right, injustice or good and bad in food production. At present there are roughly four main research areas within food ethics, presented in Table 3.2.

Table 3.2 Research areas in food ethics (Coff et al. 2008b)

a. Research areas in food ethics

1. *Food security* deals with the just and fair supply of food to human beings. With more than 900 million starving or undernourished people in the world, this is probably the most pressing ethical question
2. *Food safety* deals with the safety of the food: food should not endanger the health of consumers due to pathogens or pollution present in the food. There are ongoing discussions about what is safe enough and whose definition of safety should be followed
3. New developments in *nutritional research* and technology, such as personalised nutrition, functional foods and health foods, challenge existing norms and values about food. This also includes food-related diseases such as obesity, cardio-vascular diseases and cancer and their association with food culture, because they raise issues of responsibility and respect for "non-healthy" life-styles and production methods
4. *Ethical questions raised by specific production practises and conditions in the food chain.* This concerns animal welfare, the environment, sustainability, working conditions, use of new (bio- and nano-) technology, research ethics, and so forth. These ethics relate to the production history of the food, that is how and under what conditions it was produced

As traceability is about keeping track of the history of the food, *ethical traceability* is about keeping track of the ethical aspects of food production practices and the conditions under which the food is produced. Ethical traceability is a means of capturing and mapping values and processes in the food production chain. It can be used as a verification process of the methods and practices used, in response to consumers' ethical concerns. It can be defined in the following way:

Ethical traceability is the ability to trace and map ethical aspects of the food chain by means of recorded identifications.

Once the information on the ethical aspects of production practices has been captured and mapped, it can be used to communicate with interested stakeholders in the food chain, including producers, processors, retailers and consumers. It can be used as part of the "value-capture" of products and also to enable stakeholders to make choices consistent with their own values. Tim Lang discerns a movement away from "value for money", the idea that price is the fundamental determinant of choice, towards "values for money", reflecting the notion that consumers are more than just wallets on legs, but are also citizens who will select and reward companies that behave in socially responsible ways (Lang 2007).

3.4 Informed Food Choice

Ethical consumption mixes the role of consumer with that of citizen. The term "consumer-citizen" refers to this duality (see for instance the Consumer Citizenship Network and Scammell 2003). Clive Barnett and his colleagues (2005) consider consumer-oriented activism as a pathway to participation for ordinary people. Ethical consumption is a reconfiguration of the consumer's role, merging it with the citizen's role. From a recent attitudinal survey we know that in Europe 60% of the population is worried about animal welfare when prompted (European Commission 2006, p. 28).

Empirical sociology and psychology have taught us that there is often a gap between people's attitudes and their behaviour. Consumers' concerns as measured in surveys do rarely translate into actual food purchasing behaviour. This means that even if consumers express concerns about animal welfare (attitude) they will not necessarily purchase meat selectively (behaviour). Many factors contribute to this gap. For instance, it can be difficult to shop according to individual values due to a lack of reliable information or a lack of trust in the food system. Economic constraints or lack of easy access to food with the desired ethical attributes can also act as barriers. In Fig. 3.1 some of the major issues that influence consumers' food choice are listed (in a simplified form, as there may be many more). From the figure it is clear that choosing food is not a simple matter, and that many different and opposing interests must be weighed against each other. The priority given to different interests may vary over time, depending on the situation, the mood of the consumer or the social contexts at the moment of shopping.

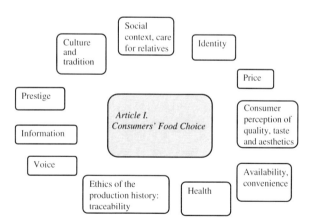

Fig. 3.1 Some factors affecting consumers' food choice (Coff et al. 2008b)

Some of the issues in Fig. 3.1 are related to self-interest. Price, quality, taste, prestige, health and convenience can all be part of self-interested considerations around the "best buy". Purely self-interested considerations can, in fact, be said to lack any ethical reflection and awareness, since they entail no concern for others.

The ethical dimension of food purchase is opened when food is bought not only for one's own satisfaction, but also to take into consideration the needs of others.

Other issues in Fig. 3.1 are related to sociological and cultural aspects of food production and consumption. For instance, it is clear that food choice is linked to culture, social class and tradition. The selection of food – the matter or "environment" to be incorporated in one's own body – confers identity not only through the social context in which it takes place but also, on a more individual level, through selection of particular foods, such as the avoidance of meat, preferences for organic food or animal-friendly meat, one's own particular preferences or dislikes, the avoidance of certain food ingredients because of their association with certain diseases, and so forth.

Information plays a crucial role for most of the issues in Fig. 3.1, and for some of the issues it is paramount. It is well documented that many animals instinctively know which plants can cure diseases and also which plants/animals should be avoided. This is no longer the case for human beings: we need knowledge to help us distinguish what is edible. In fact human food is embedded in a culture of knowledge. Food in its different social contexts relies heavily on knowledge, not so much about the food itself but on cultural traditions and habits.

However, consumers differ about which information they see as relevant; much of the information that is provided simply goes unnoticed because it is not relevant to the consumer's purposes. To be sure, in order to estimate the impact of food intake on health consumers need to be informed; but consumers have very different conceptions of health. Furthermore, information is essential for consumer decision-making as it allows for comparison between alternatives. The aim of making a comparison between different foods is to arrive at a judgement about which is the best food. "Best" depends, of course, on what criteria one considers most important. Such a judgement cannot be made without information.

3.5 Ethical Traceability in Practice

In order to explore how consumers react to the notion of ethical traceability in practice a sociological survey was set up. The idea was to analyse how people would react to an increase in information on the production history of certain articles available inside the supermarket (Coff et al. 2005). For the purpose information material on the production histories of selected articles was compiled and placed in six supermarkets, using three methods of communication:

- Computers with a homepage of pictures and print were installed in every supermarket.
- Folders using pictures and print were readily available.
- Demonstrators were present who were versed in the production history of the articles and able to inform customers about them.

This means that there were two shops with computers, two with folders and two with demonstrators. The two products used in the survey were rye bread and liver paté – two Danish staples with a large turnover in supermarkets. From the producers information on the following issues was requested:

- *Geographical origin*
- *Product quality*: Description of manufacturing methods and their effect on the quality of the product (for instance post-harvest treatments and processing techniques).
- *Production and environment*: Description of environmental initiatives and possible certifications of the product.
- *Production and animal welfare:* How are animals treated?
- *Production and working conditions*: Has anything been done in particular to ensure good working conditions?
- *Production and society:* Description of farms and businesses that are involved.
- *Economic transparency:* Who earns what? Economic fairness as an ethical parameter.
- *Guarantees and certifications:* Who is responsible for ensuring that the information on the article is correct?

However, some producers were unwilling to provide us with the required information; one of these, Danish Crown, has since been the subject of media attention for allegedly employing underpaid Polish workers in its German factories.

It also appeared that some of the actual questions were problematic. The concepts of economic transparency and fairness had to be dropped completely, since there were far too many factors at work. For example in the case of liver paté, the question of who earns what is highly relevant, since many customers suspect the major producers of squeezing out the small ones. It is easy to calculate the supermarkets' and the producers' economic share, but what about the farmer? Farmers do not sell liver; they sell whole animals from which the liver is extracted at the slaughterhouses and sold separately to liver paté producers. The slaughterhouses pay the farmers for whole animals and not separate parts, so nobody could reasonably calculate what the farmer earns for the liver. Moreover incomes must be seen in relation to the expenses of the individual parts. Large incomes in themselves are not at issue when accompanied by large expenses. Economic fairness must thus be developed in a quite different way if we are to make sense of production history.

For those businesses that wished to join the survey there were major differences in the information they possessed. In general, producers of organic products provided the most information, which is not surprising since they are subject to a voluntary market arrangement, which demands a certain degree of documentation. But a number of businesses found it impossible to answer the questions; they could only state that current legislation on animal welfare, the environment and working conditions were being observed. So a problem appeared: the letter of the law is hardly the best message to mediate. It seemed that the businesses barely had a story to tell about their products, as these were so anonymous. And in the case of the most industrialised products consumers could learn nothing about the origin or growth of the animals, nor about working conditions at the farms or the manufacturers, nor about their environmental concerns.

A third problem turned out to be the widespread market dominance by a few businesses. Three different brands of rye-bread were represented in our survey, all of which turned out to be based on flour from the same miller, Cerealia Mills. The market for flour in Denmark is dominated by three large businesses, Cerealia being the biggest. This presented two problems for the survey. Firstly, the flour could not be traced further back than to the mill where all the grain was mixed – from seven different European countries. Nor was it possible to trace the farm in question, the region, or the country of origin. This was a pity, since geographical origin is important for many people, who link a given foodstuff to a particular geographical region, landscape or culture. Secondly, this made the production histories very similar. What appeared to customers to be three different brands of rye bread were in fact very much alike. Indeed, in one of the supermarket chains the 10–12 different brands all stemmed from the same bread producer thereby partly defeating our purpose: if there is no difference in production history, only price and taste remain as delineators and ethics are irrelevant. Since businesses are not geared to distinguishing and recording production histories, they cannot provide relevant information. As a consequence our own material was not as detailed as desired. Figure 3.2 shows an example of how the production history of bread was mediated via the internet.

The production histories presented were brief in order not to burden customers more than necessary. Below is the history of Hedebagerens Rugbrød (Heathbaker's Rye Bread) from the internet page presented in Fig. 3.2.

The farm: The rye is *cultivated* by ecological growers in Denmark and abroad. This means that no synthetic pesticides or fertilisers have been used, only mechanical weeding, seed changes to avoid plant diseases and animal and "green" manure.

The farmers deliver direct to the miller (Cerealia Mills), who can trace the grain to the individual farm. However, once the ryes are mixed and ground they can no longer be traced back to the individual farms, so there is no information about working conditions on the farms. The farmers sign contracts with Cerealia Mills for delivery of the rye, which is traded at market price.

The mill: The rye is ground at Drabæksmølle (Drabaeks Mill) in Lunderskov in south-central Jutland. The rye is a 100% ground full-grain product.

Cerealia Mills is ISO 14001 certified. It identifies all environmental effects and implements measures to improve conditions by: (1) making correct environmental choices and processing changes; (2) minimising discharge by reducing the use of water; (3) minimising and removing waste in the most environmentally friendly way. Cerealia Mills aims to create a high level of job satisfaction among its staff. It is the responsibility of both the business and the staff to develop skills and knowledge.

The bakery: Hedebagerens Rye Bread is made by the bakery Le Blé d'Or (The Corn of Gold) in Amager, Copenhagen. This is a wholesale concern.

The bread is a traditional Jutland rye bread containing organic rye flour and kernels, water, leaven (sourdough), salt, and rye bread crumbs. The bread rises for at

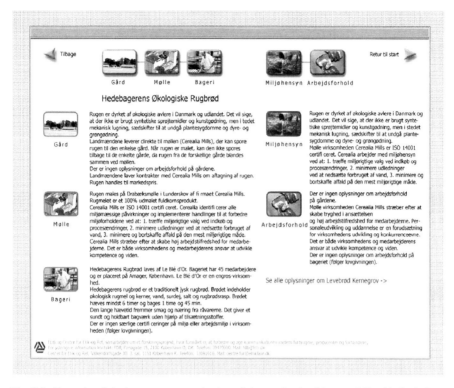

Fig. 3.2 Example of the internet communication of the production history of Heathbaker's Rye Bread. In Danish (Coff et al. 2005)

least 6 hours and is baked for 1 hour and 45 minutes. The long bread-rise improves the taste and nourishment levels of the raw materials, producing a healthy, durable bread without additives. By law no special certification is required as regards the environment or working conditions in the bakery.

The text shows how difficult it is to place a specifically ethical angle on the history of the productions. This is because only a few of the businesses involved had an explicit attitude to the ethical question.

3.6 Survey Results

In the supermarkets interviews were carried out with the shoppers in which they were asked about their opinion of the information material. Interest in the experiment was considerable: out of 257 questioned, 155 were willing to participate, 102 were not. The vast majority of the 155 participants were grateful for the increased information despite the shortcomings and the many similarities, for they knew next to nothing about production history. A number of them had a vague idea of the process, but when confronted with the material, they were often surprised; the industrialisation of food production is more widespread than most people imagine. One informant said that this form of information was *motivating*:

> I think it's fine that the information is there. Especially if like me you're not very good at finding out about it – so it motivates you.

The customer does not add in what way the information is motivating but we can imagine that there are several possibilities including the motivation to *seek* information and to *act* on the basis of ethical considerations.

Table 3.3 Main results on information, transparency, trust and ethics from empirical survey on ethical traceability (Coff et al. 2005)

Information and transparency
- Information on production history is clearly *new information* for most respondents
- Customers will very rarely be able to find information on ethical traceability but *require help and guidance*
- It is often mentioned that there is over-information on the foodstuffs market. This survey points rather to *lack of information* as an equally possible problem: that the available information is not what is being requested
- *A branding system is used for swift information* from day to day, but information as to what it stands for is not readily available; more information is required on the different brands
- Information on foodstuffs and their ethical traceability must be *easily available* and not consist of lengthy, heavy texts. It should also be provided *at the point of sale*
- For many customers *sensuality* towards the foodstuffs plays a role, but they are often excluded from this opportunity in the supermarkets

Ethical traceability
- Traceability of foodstuffs is chiefly linked to the *primary place of production*, i.e. the soil and landscape from which the food derives. For many customers the farm is more important than the processing factory
- *Geographical indication* is not merely the specification of *place*, but may have political, ethical, cultural, social or environmental significance

Trust
- A number of customers have *great trust in the retail chain*. This can be explained by their lack of contact to the producers
- *Transparency*, in the form of publicly available information and criteria for the presentation of information, creates trust
- *Impartiality* among information providers is regarded as essential by practically every respondent

Ethics
- Just over half of those interviewed thought that *increased information on the production history and ethics behind the foodstuffs would change their shopping practices*
- Information on traceability and ethics has a *motivating effect* in relation to shopping with an attitude
- The customers' interest in the ethics of production history concerns in particular *animal welfare, the environment, working conditions and fair trade*
- There are great expectations that *shops* will take ethical responsibility for the merit of their articles and for information to their customers

The survey also showed that food ethics are clearly regarded as a difficult and diffuse concept. This may be due among other things to globalisation. For modern ethics is not just about the short range and what we immediately understand. It forces us to relate to the long-range consequences of our actions which take place on the other side of the globe. It is precisely here that the difficulty lies, for we do not know these consequences.

The interviews point to a number of concrete conditions, which are of importance for the communication of the ethical traceability of the foodstuffs. The main results are presented in Table 3.3.

The main conclusions of the survey are firstly that increased information on food traceability arouses considerable customer interest. Many respondents think they need help to find their way through the consumer jungle. Secondly, the introduction of the practice of ethical traceability and information on the production history of the articles cannot be expected to show immediate results among customers. Our unsurprising experience was that it takes a long time for customers at get used to such information being available and even longer to apply it to their shopping practices. Since consumption is so much a matter of choosing, it can be regarded as a learning process (Sassatelli 2001, p 95). Consumption is not automatic; it is based on experiences, exchange of opinions, surveys and so on, not to mention advertising. Because ethics since the arrival of modern consumption has been almost excluded from the market and in consequence removed to the political sphere (at the very least theoretically), it will require a major revolution to make the ethical factor an integral part of the market. So the opposition it encounters is not surprising. Customers are used to ethical topics on production practices turning up in the media, but to be presented with them directly in the shop is new to them.

Traceability proved in the survey to be a concept with more nuances than at first assumed. Geographical indication, which in itself is no more than a simple proper name, often acquires far more significance than it immediately indicates. For by virtue of the individual images customers have of specific lands and regions, many of those interviewed are influenced by the working conditions, animal welfare and the regard for nature in general. Below we jump into the middle of one of the interviews: A woman in her 50s:

Question: Things like vegetables, it doesn't often say where they're from.
Answer: Well it should do.
Question: Is that something you think is important?
Answer: Very. Extremely important. For example, if I want to boycott articles from Israel, I sometimes have a few problems there, because there are false labels on them. Sometimes they write that they're from South Africa when in fact they're from Israel. But I look at the boxes to see where they come from!
Question: But would you use traceability to do the same with new articles?
Answer: Yes, I think it's a good idea, if for no other reason then to draw greater attention to it.

Geographical indication is not just a matter of *place*; the place of origin is linked to a number of other factors mentioned in the table above: political, ethical, cultural, social and environmental and so on. The boycott of Israel is an instance, but for example it also became increasingly clear in the course of the interviews that working conditions were not regarded as a problem for Danish products, for it is believed that Danish legislation has laid down firm regulations. This indicates that geographical indication is "over-interpreted" in the sense that consumers tend to read more into the information than they are actually given. "Produced in Denmark" refers to a country with strict regulations as regards working conditions, but in reality working conditions need not be that ideal. The speed of work, for instance, is often mentioned as a stress factor for Danish workers.

The survey also shows that the primary place of production is far more important than the secondary place of processing. This can be interpreted to mean foodstuffs and traceability are first and foremost linked to the nature and the "landscape" that produces the food.

3.7 The Future of Ethical Traceability

In the last year or two traceability of food has received growing attention from producers and retailers as a tool for consumer information and communication. Traceability is used as an instrument for creating transparency and thus to communicate ethical values in the food chain to consumers. It is therefore also seen as a means to improve consumer trust.

At present especially retailers make use of traceability as a communication tool. This might seem surprising as one should expect producers to be the first to see the benefit of using traceability as an instrument to develop new ways of communication with consumers. However, there are several reasons for this development. One of the explanations is that retailers have more direct contact with consumers than producers and therefore are more exposed and sensitive to consumer demands. As the above survey shows, in general consumers do trust retailers more than producers. The move of retailers towards implementing traceability schemes for consumer communication might be a response to the awareness of the growing importance of consumer trust for retailers. Secondly, still more retailers want to brand products with their own label, which makes trust in the retailer still more important.

Food traceability schemes in the form developed for the purpose of the survey mentioned above require some effort from the side of the consumer. The communication methods used in the referred survey are probably not user friendly enough and may today already seem old fashioned. New traceability systems that make use of scanners in personal cell phones to read barcodes on food and subsequently link up to internet pages on which information on the specific food can be found may make traceability more user friendly in the future. Presently this development is most prominent in Japan. There is no doubt that the user and consumer friendliness of traceability communication is decisive for the future development and success of traceability schemes. If user friendly traceability schemes can be developed it is indeed possible that traceability systems might replace some kinds of food labelling – and function as supplements to other kinds of labelling.

Acknowledgment I want to thank David Barling, City University London, and Michiel Korthals, Wageningen University. Both are co-authors on Chapter 1 in Christian Coff, David Barling, Michiel Korthals and Thorkild Nielsen (eds) (2008a) *Ethical Traceability and Communicating Food* on which the first part of this chapter is based.

References

Bantham A, Duval J-L (2004) Connecting food chain information for food safety/security and new value. John Deere, Food Origins, New York

Barnett C, Clarke N, Cloke P, Malpass A (2005) Articulating ethics and consumption. In: Böstrom M, Andreas F, et al. (eds) Political consumerism: Its motivations, power and conditions in the Nordic countries and elsewhere. Proceedings from the 2nd International Seminar on Political Consumerism, Oslo, August 26–29, 2004, TemaNord. Nordic Council of Ministers, Copenhagen, pp. 99–112

Coff C, Christiansen WL, Mikkelsen E (2005) Forbrugere, etik og sporbarhed (Consumers, Ethics and Traceability). Report from a survey conducted by Centre for Ethics and Law and The Danish Consumers Co-operative Society (FDB)

Coff C (2006) The taste for ethics. An ethic of food consumption. The International Library of Environmental, Agricultural and Food Ethics, vol. 7. Springer, Dordrecht

Coff C, Barling D, Korthals M, Nielsen T (eds) (2008a) Ethical traceability and communicating food. Springer, Dordrecht

Coff C, Korthals M, Barling D (2008b) Ethical traceability and informed food choice. In: Coff C, Barling D, Korthals M, Nielsen T (eds) Ethical traceability and communicating food. Springer, Dordrecht

Consumer Citizenship Network. http://www.hihm.no/eway/default.aspx?pid=252. Accessed 21 February 2008

EU Standing Committee on the Food Chain and Animal Health (2004) Guidance on the implementation of articles 11, 12, 16, 17, 18, 19 and 20 of regulation (EC) no 178/2002 on general food law

European Commission (2006) Special eurobarometer: Risk issues. 238/Wave 64.1–TNS Opinion and Social

Farm Foundation (2004) Food traceability and assurance in the global food system. Farm foundation's traceability and assurance panel report

ISO (2007) Traceability in feed and food chain: General principles and basic requirements for system design and implementation. ISO 22005, International Standards Organisation, Geneva

Lang T (2007) The new order is values-for-money. Grocer 230, 27 January:29

Sassatelli R (2001) Tamed hedonism: Choice, desires and deviant pleasures. In: Gronow J, Warde A (eds) Ordinary consumption. Routledge, London

Scammell M (2003) Citizen consumers: Towards a new marketing of politics. In: Corner J, Pels D (eds) The re-styling of politics. Sage, London

Part II
GM-Food Production

Chapter 4
Ethics and Genetically Modified Foods

Gary Comstock

Abstract This article argues that three sorts of ethical considerations converge to yield a common positive answer to the question of the ethical acceptability of GM crops: (1) the rights of people in various countries to choose to adopt GM technology; (2) the balance of likely benefits over harms to consumers and the environment from GM technology; and (3) the wisdom of encouraging discovery, innovation, and careful regulation of GM technology.

4.1 Introduction

Much of the food consumed in the United States is genetically modified (GM). GM food derives from microorganisms, plants, or animals manipulated at the molecular level to have traits that farmers or consumers desire. These foods often have been produced by techniques in which "foreign" genes are inserted into the microorganisms, plants, or animals. Foreign genes are genes taken from sources other than the organism's natural parents. In other words, GM plants contain genes they would not have contained if researchers had used only traditional plant breeding methods.

Some consumer advocates object to GM foods, and sometimes they object on ethical grounds. When people oppose GM foods on ethical grounds, they typically have some reason for their opposition. We can scrutinize their reasons and, when we do so, we are practicing applied ethics. Applied ethics involves identifying peoples' arguments for various conclusions and then analyzing those arguments to determine whether the arguments support their conclusions. A critical goal here is to decide whether an argument is sound. A sound argument is one in which all the premises are true and no mistakes have been made in reasoning.

Ethically justifiable conclusions inevitably rest on two types of claims: (i) empirical claims, or factual assertions about how the world *is*, claims ideally based on

G. Comstock (✉)
Iowa State University, Raleigh, NC, USA
e-mail: gcomstock@ncsu.edu

the best available scientific observations, principles, and theories; and (ii) normative claims, or value-laden assertions about how the world *ought to be*, claims ideally based on the best available moral judgments, principles, and theories.

Is it ethically justifiable to pursue GM crops and foods? There is an objective answer to this question, and we will try here to figure out what it is. But we must begin with a proper, heavy, dose of epistemic humility, acknowledging that few ethicists at the moment seem to think they know the final answer.

Should the law allow GM foods to be grown and marketed? The answer to this, and every, public policy question rests ultimately with us, citizens who will in the voting booth and shopping market decide the answer. To make up our minds, we will use feelings, intuitions, conscience, and reason. However, as we citizens are, by and large, not scientists, we must, to one degree or other, rest our factual understanding of the matter on the opinions of scientific experts. Therefore, ethical responsibility in the decision devolves heavily on scientists engaged in the new GM technology.

4.2 Ethical Responsibilities of Scientists

Science is a communal process devoted to the discovery of knowledge and to open and honest communication of knowledge. Its success, therefore, rests on two different kinds of values.

Epistemological values are values by which scientists determine which knowledge claims are better than others. The values include clarity, objectivity, capacity to explain a range of observations, and ability to generate accurate predictions. Claims that are internally inconsistent are jettisoned in favor of claims that are consistent and fit with established theories. (At times, anomalous claims turn out to be justifiable, and an established theory is overthrown, but these occasions are rare in the history of science.) Epistemological values in science also include fecundity, the ability to generate useful new hypotheses; simplicity, the ability to explain observations with the fewest number of additional assumptions or qualifications; and elegance.

Personal values, including honesty and responsibility, are a second class of values – values that allow scientists to trust their peers' knowledge claims. If scientists are dishonest, untruthful, fraudulent, or excessively self-interested, the free flow of accurate information so essential to science will be thwarted. If a scientist plagiarizes the work of others or uses fabricated data, that scientist's work will become shrouded in suspicion and otherwise reliable data will not be trusted. If scientists exploit those who work under them or discriminate on the basis of gender, race, class, or age, then the mechanisms of trust and collegiality undergirding science will be eroded.

The very institution of scientific discovery is supported – indeed, permeated – with values. Scientists have a variety of goals and functions in society, so it should be no surprise that they face different challenges.

University scientists must be scrupulous in giving credit for their research to all who deserve credit; careful not to divulge proprietary information; and painstaking in maintaining objectivity, especially when funded by industry. Industry scientists must also maintain the highest standards of scientific objectivity, a particular challenge because their work may not be subject to peer-review procedures as strict as those faced by university scientists. Industry scientists must also be willing to defend results of their research that are not favorable to their employer's interests. Scientists employed by non-governmental organizations face challenges as well. Their objectivity must be maintained in the face of an organization's explicit advocacy agenda and in spite of the fact that their research might provide results that could seriously undermine the organization's fund-raising attempts. All scientists face the challenges of communicating complex issues to a public that receives them through media channels that often are not equipped to communicate the qualifications and uncertainties attached to much scientific information.

At its core, science is an expression of some of our most cherished values. The public largely trusts scientists, and scientists must in turn act as good stewards of this trust.

4.3 A Method for Addressing Ethical Issues

Ethical objections to GM foods typically center on the possibility of harm to persons or other living things. Harm may or may not be justified by outweighing benefits. Whether harms are justified is a question that ethicists try to answer by working methodically through a series of questions:

> What is the harm envisaged? To provide an adequate answer to this question, we must pay attention to how significant the harm or potential harm may be (will it be severe or trivial?); who the "stakeholders" are (that is, who are the persons, animals, even ecosystems, who may be harmed?); the extent to which various stakeholders might be harmed; and the distribution of harms. The last question directs attention to a critical issue, the issue of justice and fairness. Are those who are at risk of being harmed by the action in question different from those who may benefit from the action in question?
> What information do we have? Sound ethical judgments go hand in hand with a thorough understanding of the scientific facts. In a given case, we may need to ask two questions. Is the scientific information about harm being presented reliable, or is it fact, hearsay, or opinion? What information do we not know that we should know before we make the decision?
> What are the options? In assessing the various courses of action, emphasize creative problem-solving, seeking to find win-win alternatives in which everyone's interests are protected. Here we must identify what objectives each stakeholder wants to obtain; how many methods are available by which

to achieve those objectives; and what advantages and disadvantages attach to each alternative.

What ethical principles should guide us? There are at least three secular ethical traditions:

> Rights theory holds that we ought always to act so that we treat human beings as autonomous individuals and not as mere means to an end.
>
> Utilitarian theory holds that we ought always to act so that we maximize good consequences and minimize harmful consequences.
>
> Virtue theory holds that we ought always to act so that we act the way a just, fair, good person would act.
>
> Ethical theorists are divided about which of these three theories is best. We manage this uncertainty through the following procedure. Pick one of the three principles. Using it as a basis, determine its implications for the decision at hand. Then, adopt a second principle. Determine what it implies for the decision at hand. Repeat the procedure with the third principle. Should all three principles converge on the same conclusion, then we have good reasons for thinking our conclusion is morally justifiable.

How do we reach moral closure? Does the decision we have reached allow all stakeholders either to participate in the decision or to have their views represented? If a compromise solution is deemed necessary in order to manage otherwise intractable differences, has the compromise been reached in a way that has allowed all interested parties to have their interests articulated, understood, and considered? If so, then the decision may be justifiable on ethical grounds.

There is a difference between consensus and compromise. Consensus means that the vast majority of people agree about the right answer to a question. If the group cannot reach a consensus but must, nevertheless, make some decision, then a compromise position may be necessary. But neither consensus nor compromise should be confused with the right answer to an ethical question. It is possible that a society might reach a consensus position that is unjust. For example, some societies have held that women should not be allowed to own property. That may be a consensus position or even a compromise position, but it should not be confused with the truth of the matter. Moral closure is a sad fact of life; we sometimes must decide to undertake some course of action even though we know that it may not be, ethically, the right decision, all things considered.

4.4 Ethical Issues Involved in the Use of Genetic Technology in Agriculture

Discussions of the ethical dimensions of agricultural biotechnology (ag biotech) are sometimes confused by a conflation of two quite different sorts of objections to GM technology: intrinsic and extrinsic. It is critical not only that we distinguish these two classes but that we keep them distinct throughout the ensuing discussion of ethics.

Extrinsic objections focus on the potential harms consequent upon the adoption of GM organisms (GMOs). Extrinsic objections hold that GM technology should not be pursued because of its anticipated results. Briefly stated, the extrinsic objections can be described as follows. GMOs may have disastrous effects on animals, ecosystems, and humans. Possible harms to humans include perpetuation of social inequities in modern agriculture, decreased food security for women and children on subsistence farms in developing countries, a growing gap between well-capitalized economies in the northern hemisphere and less capitalized peasant economies in the South, risks to the food security of future generations, and the promotion of reductionistic and exploitative science. Potential harms to ecosystems include possible environmental catastrophe; inevitable narrowing of germplasm diversity; and irreversible loss or degradation of air, soils, and waters. Potential harms to animals include unjustified pain to individuals used in research and production.

These are valid concerns, and nation-states must have in place testing mechanisms and regulatory agencies to assess the likelihood, scope, and distribution of potential harms through a rigorous and well-funded risk assessment procedure. It is for this reason that I said above that GM technology must be developed responsibly and with appropriate caution. However, these extrinsic objections cannot by themselves justify a moratorium, much less a permanent ban, on GM technology, because they admit the possibility that the harms may be minimal and outweighed by the benefits. How can one decide whether the potential harms outweigh potential benefits unless one conducts the research, field tests, and data analysis necessary to make a scientifically informed assessment?

In sum, extrinsic objections to GMOs raise important questions about GMOs, and each country using GMOs ought to have in place the organizations and research structures necessary to ensure their safe use.

There is, however, an entirely different sort of objection to GM technology, a sort of objection that, if it is sound, would justify a permanent ban.

Intrinsic objections allege that the process of making GMOs is objectionable *in itself*. This belief is defended in several ways, but almost all the formulations are related to one central claim, the unnaturalness objection:

It is unnatural to genetically engineer plants, animals, and foods (*UE*).

If *UE* is true, then we ought not to engage in bioengineering, however unfortunate may be the consequences of halting the technology. Were a nation to accept *UE* as the conclusion of a sound argument, then much agricultural research would have to be terminated and potentially significant benefits from the technology sacrificed. A great deal is at stake.

In Comstock, *Vexing Nature? On the Ethical Case Against Agricultural Biotechnology*, I discuss 14 ways in which *UE* has been defended (Comstock 2000). For present purposes, those 14 objections can be summarized as follows:

- To engage in ag biotech is to *play God*.
- To engage in ag biotech is to invent world-changing technology.
- To engage in ag biotech is illegitimately to cross species boundaries.
- To engage in ag biotech is to *commodify life*.

Let us consider each claim in turn.

4.4.1 To Engage in ag Biotech Is to Play God

In a western theological framework, humans are creatures, subjects of the Lord of the Universe, and it would be impious for them to arrogate to themselves roles and powers appropriate only for the Creator. Shifting genes around between individuals and species is taking on a task not appropriate for us, subordinate beings. Therefore, to engage in bioengineering is to play God.

There are several problems with this argument. First, there are different interpretations of God. Absent the guidance of any specific religious tradition, it is logically possible that God could be a Being who wants to turn over to us all divine prerogatives, or explicitly wants to turn over to us at least the prerogative of engineering plants, or who doesn't care what we do. If God is any of these beings, then the argument fails because playing God in this instance is not a bad thing.

The argument seems to assume, however, that God is not like any of the gods just described. Assume that the orthodox Jewish and Christian view of God is correct, that God is the only personal, perfect, necessarily existing, all-loving, all-knowing, and all-powerful being. On this traditional western theistic view, finite humans should not aspire to infinite knowledge and power. To the extent that bioengineering is an attempt to control nature itself, the argument would go, bioengineering would be an unacceptable attempt to usurp God's dominion.

The problem with this argument is that not all traditional Jews and Christians think this God would rule out genetic engineering. I am a practicing evangelical Christian and the chair of my local church's council. In my tradition, God is thought to endorse creativity and scientific and technological development, including genetic improvement. Other traditions have similar views. In the mystical writings of the Jewish Kabbalah, God is understood as One who expects humans to be co-creators, technicians working with God to improve the world. At least one Jewish philosopher, Baruch Brody, has suggested that biotechnology may be a vehicle ordained by God for the perfection of nature (B. Brody, private communication).

I personally hesitate to think that humans can perfect nature. However, I have become convinced that GM might help humans to rectify some of the damage we have already done to nature. And I believe God may endorse such an aim. For humans are made in the divine image. God desires that we exercise the spark of divinity within us. Inquisitiveness in science is part of our nature. Creative impulses are not found only in the literary, musical, and plastic arts. They are part of molecular biology, cellular theory, ecology, and evolutionary genetics, too. It is unclear why the desire to investigate and manipulate the chemical bases of life should not be considered as much a manifestation of our god-like nature as the writing of poetry and the composition of sonatas. As a way of providing theological content for *UE*, then, argument (i) is unsatisfactory because it is ambiguous and contentious.

4.4.2 To Engage in ag Biotech Is to Invent World-Changing Technology, an Activity that Should Be Reserved to God Alone

Let us consider (ii) in conjunction with a similar objection (iia).

(iia) To engage in ag biotech is to *arrogate historically unprecedented power* to ourselves.

The argument here is not the strong one, that biotech gives us divine power, but the more modest one, that it gives us a power we have not had previously. But it would be counterintuitive to judge an action wrong simply because it has never been performed. On this view, it would have been wrong to prescribe a new herbal remedy for menstrual cramps or to administer a new anesthetic. But that seems absurd. More argumentation is needed to call historically unprecedented actions morally wrong. What is needed is to know *to what extent* our new powers will transform society, whether we have witnessed prior transformations of this sort, and whether those transitions are morally acceptable.

We do not know how extensive the ag biotech revolution will be, but let us assume that it will be as dramatic as its greatest proponents assert. Have we ever witnessed comparable transitions? The change from hunting and gathering to agriculture was an astonishing transformation. With agriculture came not only an increase in the number of humans on the globe but the first appearance of complex cultural activities: writing, philosophy, government, music, the arts, and architecture. What sort of power did people arrogate to themselves when they moved from hunting and gathering to agriculture? The power of civilization itself (McNeill 1989).

Ag biotech is often oversold by its proponents. But suppose they are right, that ag biotech brings us historically unprecedented powers. Is this a reason to oppose it? Not if we accept agriculture and its accompanying advances, for when we accepted agriculture we arrogated to ourselves historically unprecedented powers.

In sum, the objections stated in (ii) and (iia) are not convincing.

4.4.3 To Engage in ag Biotech Is Illegitimately to Cross Species Boundaries

The problems with this argument are both theological and scientific. I will leave it to others to argue the scientific case that nature gives ample evidence of generally fluid boundaries between species. The argument assumes that species boundaries are distinct, rigid, and unchanging, but, in fact, species now appear to be messy, plastic, and mutable. To proscribe the crossing of species borders on the grounds that it is unnatural seems scientifically indefensible.

It is also difficult to see how (iii) could be defended on theological grounds. None of the scriptural writings of the western religions proscribes genetic engineering, of course, because genetic engineering was undreamt of when the holy books were written. Now, one might argue that such a proscription may be derived from Jewish or Christian traditions of scriptural interpretation. Talmudic laws against mixing "kinds," for example, might be taken to ground a general prohibition against inserting genes from "unclean" species into clean species. Here's one way the argument might go: For an observant Jew to do what scripture proscribes is morally wrong; Jewish oral and written law proscribe the mixing of kinds (eating milk and meat from the same plate; yoking donkeys and oxen together); bioengineering is the mixing of kinds; therefore, for a Jew to engage in bioengineering is morally wrong.

But this argument fails to show that bioengineering is intrinsically objectionable in all its forms for everyone. The argument might prohibit *Jews* from engaging in certain *kinds* of biotechnical activity but not all; it would not prohibit, for example, the transferring of genes *within* a species, nor, apparently, the transfer of genes from one clean species to another clean species. Incidentally, it is worth noting that the Orthodox community has accepted transgenesis in its food supply. Seventy percent of cheese produced in the United States is made with a GM product, chymosin . This cheese has been accepted as kosher by Orthodox rabbis (Gressel 1999).

In conclusion, it is difficult to find a persuasive defense of (iii) on either scientific or religious grounds.

4.4.4 To Engage in ag Biotech Is to Commodify Life

The argument here is that genetic engineering treats life in a reductionistic manner, reducing living organisms to little more than machines. Life is sacred and not to be treated as a good of commercial value only to be bought and sold to the highest bidder.

Could we apply this principle uniformly? Would not objecting to the products of GM technology on these grounds also require that we object to the products of ordinary agriculture on the same grounds? Is not the very act of bartering or exchanging crops and animals for cash vivid testimony to the fact that every culture on earth has engaged in the commodification of life for centuries? If one accepts commercial trafficking in non-GM wheat and pigs, then why object to commercial trafficking in GM wheat and GM pigs? Why should it be wrong to treat DNA the way we have previously treated animals, plants, and viruses (Nelkin and Lindee 1995)?

Although (iv) may be true, it is not a sufficient reason to object to GM technology because our values and economic institutions have long accepted the commodification of life. Now, one might object that various religious traditions have never accepted commodification and that genetic engineering presents us with an opportunity to resist, to reverse course. Kass (1988, HN1), for example, has argued that we have gone too far down the road of dehumanizing ourselves and treating nature as a machine and that we should pay attention to our emotional reactions against practices such as human cloning. Even if we cannot defend these feelings in rational terms, our revulsion at the very idea of cloning humans should carry great weight. Midgley (2000) has argued that moving genes across species boundaries is not only "yukky" but, perhaps, a monstrous idea, a form of playing God.

Kass and Midgley have eloquently defended the relevance of our emotional reactions to genetic engineering but, as both admit, we cannot simply allow our emotions to carry the day. As Midgley writes, "Attention to ... sympathetic feelings [can stir] up reasoning that [alters] people's whole world view" (p. 10). But as much hinges on the reasoning as on the emotions.

Are the intrinsic objections sound? Are they clear, consistent, and logical? Do they rely on principles we are willing to apply uniformly to other parts of our lives? Might they lead to counterintuitive results?

Counterintuitive results are results we strongly hesitate to accept because they run counter to widely shared considered moral intuitions. If a moral rule or principle leads to counterintuitive results, then we have a strong reason to reject it. For example, consider the following moral principle, which we might call the doctrine of naive consequentialism (NC):

Always improve the welfare of the most people (NC).

Were we to adopt NC, then we would be not only permitted but required to sacrifice one healthy person if by doing so we could save many others. If six people need organ transplants (two need kidneys, one needs a liver, one needs a heart, and two need lungs) then NC instructs us to sacrifice the life of the healthy person to transplant six organs to the other six. But this result, that we are *obliged* to sacrifice innocent people to save strangers, is wildly counterintuitive. This result gives us a strong reason to reject NC.

I have argued that the four formulations of the unnaturalness objection considered above are unsound insofar as they lead to counterintuitive results. I do not take this position lightly. Twelve years ago, I wrote "The Case Against bGH," an article, I have been told, that "was one of the first papers by a philosopher to object to ag biotech on explicitly ethical grounds." I then wrote a series of other articles objecting to GM herbicide-resistant crops, transgenic animals, and, indeed, all of ag biotech (see Comstock 1988). I am acquainted with worries about GM foods. But, for reasons that include the weakness of the intrinsic objections, I have changed my mind. The sympathetic feelings on which my anti-GMO worldview was based did not survive the stirring up of reasoning.

4.5 Why Are We Careful with GM Foods?

I do not pretend to know anything like the full answer to this question, but I would like to be permitted the luxury of a brief speculation about it.

The reason may have to do with a natural, completely understandable, and wholly rational tendency to take precautions with what goes into our mouths. When we are in good health and happy with the foods available to us, we have little to gain from experimenting with new food and no reason to take a chance on a potentially unsafe food. We may think of this disposition as the precautionary response.

When faced with two contrasting opinions about issues related to food safety, consumers place great emphasis on negative information. The precautionary response is particularly strong when a consumer sees little to gain from a new food technology.

When a given food is plentiful, it is rational to place extra weight on negative information about any particular piece of that food. It is rational to do so, as my colleague Dermot Hayes points out, even when the source of the negative information is known to be biased.

There are several reasons for us to take a precautionary approach to new foods. First, under conditions in which nutritious tasty food is plentiful, we have nothing to gain from trying a new food if, from our perspective, it is in other respects identical to our current foods. Suppose on a rack in front of me there are 18 dozen maple-frosted Krispy Kreme doughnuts, all baked to a golden brown, all weighing three ounces. If I am invited to take one of them, I have no reason to favor one over the other.

Suppose, however, that a naked man runs into the room with wild-hair flying behind him yelling that the sky is falling. He approaches the rack and points at the third doughnut from the left on the fourth shelf from the bottom. He exclaims, "This doughnut will cause cancer! Avoid it at all costs, or die!" There is no reason to believe this man's claim and yet, because there are so many doughnuts freely available, why should we take any chances? It is rational to select other doughnuts, because all are alike. Now, perhaps one of us is a mountain climber who loves taking risks and might be tempted to say, "Heck, I'll try that doughnut." In order to focus on the right question here, the risk takers should ask themselves whether they would select the tainted doughnut to take home to feed to their 2-year-old daughter. Why impose any risk on your loved ones when there is no reason to do so?

The Krispy Kreme example is meant to suggest that food tainting is both a powerful and an extraordinarily easy social act. It is powerful because it virtually determines consumer behavior. It is easy, because the tainter does not have to offer any evidence of the food's danger. Under conditions of food plenty, rational consumers do and should take precautions, avoiding tainted food no matter how untrustworthy the tainter.

Our tendency to take precautions with our food suggests that a single person with a negative view about GM foods will be much more influential than many people with a positive view. The following experiment lends credibility to this hypothesis. In a willingness-to-pay experiment, Hayes and colleagues paid 87 primary food shoppers $40 each (Fox et al. 2002). Each participant was assigned to a group ranging in size from a half-dozen to a dozen members. Each group was then seated at a table at lunchtime and given one pork sandwich. In the middle of each table was one additional food item, an irradiated pork sandwich. Each group of participants was given one of three different treatments: (i) the *pro-irradiation* treatment; (ii) the *anti-irradiation* treatment; and (iii) the *balanced* treatment.

Each treatment began with all the participants at a table receiving the same, so-called "neutral" description of an irradiated pork sandwich. The description read, in part, like this:

The U.S. FDA has recently approved the use of ionizing radiation to control Trichinella in pork products. This process results in a 10,000-fold reduction in Trichinella organisms in meat. The process does not induce measurable radioactivity in food.

After the participants read this description, they conducted a silent bid to purchase the right to exchange their nonirradiated sandwich for the irradiated sandwich. Whoever bid the highest price would be able to buy the sandwich for the price bid by the second-highest bidder. To provide participants with information about the opinions of the others at their table so that they could factor this information into their future bids, the lowest and highest bids of each round were announced before the next round of bidding began. At the end of the experiment, 1 of the 10 bidding rounds was selected at random, and the person bidding the highest amount in that round had to pay the second-highest price bid during that round for the sandwich.

After five rounds of bidding, the second-highest bids in all three groups settled rather quickly at an equilibrium point, roughly 20 cents. That is, someone at every table was willing to pay 20 cents for the irradiated pork sandwich, but no one in any group would pay more than 20 cents. The bidding was repeated five times in order to give participants the opportunity to respond to information they were getting from others at the table and to ensure the robustness of the price.

After five rounds of bidding, each group was given additional information. Group a, the so-called *pro* group, was provided with a description of the sandwich that read, in part: Each year, 9,000 people die in the United States from food-borne illness. Some die from Trichinella in pork. Millions of others suffer short-term illness. Irradiated pork is a safe and reliable way to eliminate this pathogen. The process has been used successfully in 20 countries since 1950.

The *pro*-group participants were informed that the author of this positive description was a pro-irradiation food industry group. After the description was read, five more rounds of bidding began. The price of the irradiated sandwich quickly shot upward, reaching 80 cents by the end of round 10. A ceiling price was not reached, however, as the bids in every round, including the last, were significantly higher than in the preceding round. The price, that is, was still going up when the experiment was stopped (see Fig. 4.1).

After its first five rounds of bidding, group b was provided with a different description. It read, in part:

In food irradiation, pork is exposed to radioactive materials. It receives 300,000 rads of radiation – the equivalent of 30 million chest X-rays. This process results in radiolytic products in food. Some radiolytic products are carcinogens and linked to birth defects. The process was developed in the 1950s by the Atomic Energy Commission.

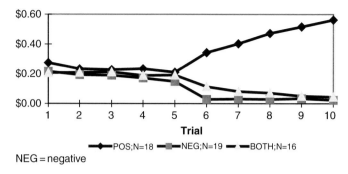

Fig. 4.1 Effect of information on average bid for irradiated pork: POS, positive; NEG, negative

The source of this description was identified to the bidders as "Food and Water," an anti-irradiation activist group in England. After group b read this description, it began five more rounds of bidding. The bid went down, quickly reaching zero. After the first five rounds produced a value of 20 cents in group b for the pork sandwich described in a "neutral" way, *no one* in this group would pay a penny for the irradiated sandwich described in a "negative" way. This result was obtained even though the description was clearly identified as coming from an activist, nonscientific group.

After five rounds of bidding on the neutral description, the third group, group c, received *both* the positive and negative descriptions. One might expect this group's response to be highly variable, with some participants scared off by the negative description and others discounting it for its unscientific source. Some participants might be expected to bid nothing while others would continue to bid highly.

However, the price of the sandwich in the third, so-called *balanced* group, also fell quickly. Indeed, the price reached zero almost as quickly as it did in group b, the negative group. That is, even though the third group had both the neutral and the positive description in front of them, no one exposed to the negative description would pay 2 cents for the irradiated sandwich.

Hayes' study illuminates the precautionary response and carries implications for the GM debate. These implications are that, given neutral or positive descriptions of GM foods, consumers initially will *pay more* for them. Given negative descriptions of GM foods, consumers initially will *not* pay more for them. Finally, and this is the surprising result, given *both* positive and negative descriptions of GM foods, consumers initially will *not* pay more for them. Both sides in the GM food debate should be scrupulous in providing reasons for all their claims. But especially for their negative claims.

In a worldwide context, the precautionary response of those facing food abundance in developed countries may lead us to be insensitive to the conditions of those in less fortunate situations. Indeed, we may find ourselves in the following ethical dilemma.

For purposes of argument, make the following three assumptions. (I do not believe any of the assumptions is implausible.) First, assume that GM food is safe. Second, assume that some GM "orphan" foods, such as rice enhanced with iron or vitamin A, or virus-resistant cassava, or aluminum-tolerant sweet potato, may be of great potential benefit to millions of poor children. Third, assume that widespread anti-GM information and sentiment, no matter how unreliable on scientific grounds, could shut down the GM infrastructure in the developed world.

Under these assumptions, consider the possibility that, by tainting GM foods in the countries best suited to conduct GM research safely, anti-GM activists could bring to a halt the range of money-making GM foods marketed by multinational corporations. This result might be a good or a bad thing. However, an unintended side effect of this consequence would be that the new GM orphan crops mentioned above might not be forthcoming, assuming that the development and commercialization of these orphan crops depends on the answering of fundamental questions in plant science and molecular biology that will be answered only if the research agendas of private industry are allowed to go forward along with the research agendas of public research institutions.

Our precautionary response to new food may put us in an uncomfortable position. On the one hand, we want to tell "both sides" of the GM story, letting people know about the benefits and the risks of the technology. On the other hand, some of the people touting the benefits of the technology make outlandish claims that it will feed the world and some of the people decrying the technology make unsupported claims that it will ruin the world. In that situation, however, those with unsupported negative stories to tell carry greater weight than those with unsupported positive stories. Our precautionary response, then, may well lead, in the short term at least, to the rejection of GM technology. Yet, the rejection of GM technology could indirectly harm those children most in need, those who need what I have called the orphan crops. Are we being forced to choose between two fundamental values, the value of free speech versus the value of children's lives?

On the one hand, open conversation and transparent decision-making processes are critical to the foundations of a liberal democratic society. We must reach out to include everyone in the debate and allow people to state their opinions about GM foods, whatever their opinion happens to be, whatever their level of acquaintance with the science and technology happens to be. Free speech is a value not to be compromised lightly.

On the other hand, stating some opinions about GM food can clearly have a tainting effect, a powerful and extraordinarily easy consequence of free speech. Tainting the technology might result in the loss of this potentially useful tool. Should we, then, draw some boundaries around the conversation, insisting that each contributor bring some measure of scientific data to the table, especially when negative claims are being made? Or are we collectively prepared to leave the conversation wide open? That is, in the name of protecting free speech, are we prepared to risk losing an opportunity to help some of the world's most vulnerable?

4.6 The Precautionary Principle

As a 13 year-old, I won my dream job, wrangling horses at Honey Rock Camp in northern Wisconsin. The image I cultivated for myself was the weathered cowboy astride Chief or Big Red, dispensing nuggets to awestruck young rider wannabes. But I was, as they say in Texas, all hat.

"Be careful!" was the best advice I could muster.

Only after years of experience in a western saddle would I have the skills to size up various riders and advise them properly on a case-by-case basis. You should slouch more against the cantle and get the balls of your feet onto the stirrups. You need to thrust your heels in front of your knees and down toward the animal's front hooves. You! Roll your hips in rhythm with the animal, and stay away from the horn. You, stay alert for sudden changes of direction.

Only after years of experience with hundreds of different riders would I realize that my earlier generic advice, well-intentioned though it was, had been of absolutely no use to anyone. As an older cowboy once remarked, I might as well have been saying, "Go crazy!" Both pieces of advice were equally useless in making good decisions about how to behave on a horse.

Now, as mad cow disease grips the European imagination, concerned observers transfer fears to genetically modified foods, advising: "Take precaution!" Is this a valuable observation that can guide specific public policy decisions, or well-intentioned but ultimately unhelpful advice?

As formulated in the 1992 Rio Declaration on Environment and Development, the precautionary principle states that "... lack of full scientific certainty shall not be used as a reason for postponing cost-effective measures to prevent environmental degradation." The precautionary approach has led many countries to declare a moratorium on GM crops on the supposition that developing GM crops might lead to environmental degradation. The countries are correct that this is an implication of the principle. But is it the only implication?

Suppose global warming intensifies and comes, as some now darkly predict, to interfere dramatically with food production and distribution. Massive dislocations in international trade and corresponding political power follow global food shortages, affecting all regions and nations. In desperate attempts to feed themselves, billions begin to pillage game animals, clear-cut forests to plant crops, cultivate previously nonproductive lands, apply fertilizers and pesticides at higher than recommended rates, kill and eat endangered and previously nonendangered species.

Perhaps not a likely scenario, but not entirely implausible, either. GM crops could help to prevent it, by providing hardier versions of traditional lines capable of growing in drought conditions, or in saline soils, or under unusual climactic stresses in previously temperate zones, or in zones in which we have no prior agronomic experience.

On the supposition that we might need the tools of genetic engineering to avert future episodes of crushing human attacks on what Aldo Leopold called "the land", the precautionary principle requires that we develop GM crops. Yes, we lack full scientific certainty that developing GM crops will prevent environmental degradation. True, we do not know what the final financial price of GM research and development will be. But if GM technology were to help save the land, few would not deem that price cost-effective. So, according to the precautionary principle, lack of full scientific certainty that GM crops will prevent environmental degradation shall not be used as a reason for postponing this potentially cost-effective measure.

The precautionary principle commits us to each of the following propositions:

(1) We must not develop GM crops.
(2) We must develop GM crops.

As (1) and (2) are plainly contradictory, however, defenders of the principle should explain why its implications are not incoherent.

Much more helpful than the precautionary principle would be detailed case-by-case recommendations crafted upon the basis of a wide review of nonindustry-sponsored field tests conducted by objective scientists expert in the construction and interpretation of ecological and medical data. Without such a basis for judging this use acceptable and that use unacceptable, we may as well advise people in the GM area to go crazy. It would be just as helpful as "Take precaution!"

4.7 Religion and Ethics

Religious traditions provide an answer to the question, "How, overall, should I live my life?" Secular ethical traditions provide an answer to the question, "What is the right thing to do?" When in a pluralistic society a particular religion's answers come into genuine conflict with the answers arrived at through secular ethical deliberation, we must ask how deep is the conflict. If the conflict is so deep that honoring the religion's views would entail dishonoring another religion's views, then we have a difficult decision to make. In such cases, the conclusions of secular ethical deliberation must override the answers of the religion in question.

The reason is that granting privileged status to one religion will inevitably discriminate against another religion. Individuals must be allowed to follow their conscience in matters theological. But if one religion is allowed to enforce its values on others in a way that restricts the others' ability to pursue their values, then individual religious freedom has not been protected.

Moral theorists refer to this feature of nonreligious ethical deliberation as the *overridingness* of ethics. If a parent refuses a life-saving medical procedure for a minor child on religious grounds, the state is justified in overriding the parent's religious beliefs in order to protect what secular ethics regards as a value higher than religious freedom: the life of a child.

The overridingness of ethics applies to our discussion only if a religious group claims the right to halt GM technology on purely religious grounds. The problem here is the confessional problem of one group attempting to enforce its beliefs on others. I mean no disrespect to religion; as I have noted, I am a religious person, and I value religious traditions other than my own. Religious traditions have been the repositories and incubators of virtuous behavior. Yet each of our traditions must in a global society learn to coexist peacefully with competing religions and with nonreligious traditions and institutions.

If someone objects to GM technology on purely religious grounds, we must ask on what authority they speak for their tradition, whether there are other, conflicting, views within their tradition and whether acting on their views will entail disrespecting the views of people from other religions. It is, of course, the right of each tradition to decide its attitude about genetic engineering. But in the absence of other good reasons, we must not allow someone to ban GM technology for narrowly sectarian reasons alone. To allow such an action would be to disrespect the views of people who believe, on equally sincere religious grounds, that GM technology is not necessarily inconsistent with God's desires for us.

4.8 Minority Views

When in a pluralistic society the views of a particular minority come into genuine conflict with the views of the majority, we must ask a number of questions: How deep is the conflict? How has the minority been treated in the past? If the minority has been exploited, have reparations been made? If the conflict is so deep that honoring the minority's views would entail overriding the majority's views, then we have a difficult decision to make. In such cases, the conclusions of the state must be just, taking into account the question of past exploitation and subsequent reparations or lack thereof. This is a question of justice.

The question of justice would arise in the discussion of GM technology if the majority favored GM technology, and the minority claimed the right to halt GM technology. If the minority cites religious arguments to halt GMOs and the majority believes that halting GMOs will result in loss of human life, then the state faces a decision very similar to the one discussed in the prior section. In this case, secular policy decisions may be justified in overriding the minority's religious arguments insofar as society deems the value of human life higher than the value of religious freedom.

However, should the minority cite past oppression as the reason their values ought to predominate over the majority's, then a different question must be addressed. Here, the relevant issues have to do with the nature of past exploitation, its scope and depth, and the sufficiency of efforts, have there been any, to rectify the injustice and compensate victims. If the problem is long-standing and has not been addressed, then imposing the will of the majority would seem a sign of an unjust society insensitive to its past misdeeds. If, on the other hand, the problem has been carefully addressed by both sides and, for example, just treaties arrived at through fair procedures have been put in place, are being enforced, are rectifying past wrongs, and are preventing new forms of exploitation, then the minority's arguments would seem to be far weaker. This conclusion would be especially compelling if it could be shown that the lives of *other* disadvantaged peoples might be put at risk by honoring a particular minority's wish to ban GMOs.

4.9 Conclusion

Earlier I described a method for reaching ethically sound judgments. It was on the basis of that method that I personally came to change my mind about the moral acceptability of GM crops. My opinion changed as I took full account of three considerations: (i) the rights of people in various countries to choose to adopt GM technology (a consideration falling under the human rights principle); (ii) the balance of likely benefits over harms to consumers and the environment from GM technology (a utilitarian consideration); and (iii) the wisdom of encouraging discovery, innovation, and careful regulation of GM technology (a consideration related to virtue theory).

Is it ethically justifiable to pursue GM crops and foods? I have come to believe that three of our most influential ethical traditions converge on a common answer. Assuming we proceed responsibly and with appropriate caution, the answer is yes.

Acknowledgment Parts of this chapter were previously published:

Comstock G (2000) Vexing nature? On the ethical case against agricultural biotechnology. Kluwer Academic Publishers, Boston/Dordrecht, pp. 182–195, reprinted with permission of the publisher.

Comstock G (2001) Ethics and genetically modified foods, reprinted with permission of SCOPE GM Food Controversy Forum (1 July 2001). Copyright ©2001 American Association for the Advancement of Science.

Comstock G (2000) Make plans on the hoof. In: Times Higher Education Supplement (London) 22–29 Dec 2000, p. 19.

I learned much from discussing these ideas with colleagues, especially G. Varner, T. Smith, N. Hettinger, M. Saner, R. Streiffer, D. Hayes, K. Hessler, F. Kirschenmann, and C.S. Prakash. I was also fortunate to participate in several conversations on the topic during the past few months, and would like to express gratitude to my hosts, including the following:

Three local chapters of the American Chemical Society at Eastern Oregon University (R. Hermens), Washington State University (R. Willett), and Seattle University (S. Jackels) in Oct 2000.

The "New Zealand Royal Commission on Genetic Modification"; a public audience in Wellington, New Zealand (sponsored by the New Zealand Life Sciences Network, and F Wevers); and St. John's College, Auckland, New Zealand (G. Redding), Nov 2000.

The "Plant Sciences Institute Colloquium", Iowa State University, Feb 2001 (S. Howell).

"Biotech Issues 2001", an Extension In-Service conference at Colorado State University (B. Zimdahl and P. Kendall); and a seminar in the CSU Philosophy Department (P. Cafaro and H. Rolston); both in Feb 2001.

The 2001 Annual Meeting of the American Association for the Advancement of Science, San Francisco, in February (K.R. Smith and N. Ballenger).

A seminar at the University del Pais Vasco/Euskal Herriko Unib., Vitoria, Spain, in March (M. Salona, M. de Renobales).

A seminar in the Departamento de Microbiologia e Instituto de Biotecnología, Universidad de Granada, Spain, in March (E. Ianez).

A colloquium on environmental ethics, "Colóquio Ética Ambiental: uma ética para o futuro" at Faculdade de Letras da Universidade de Lisboa, Mar 2000 (C. Beckert).

A seminar at the Center for International Development and Science, the Technology and Public Policy Program, and the Belfer Center for Science and International Affairs, Harvard University, Mar 2001 (C. Juma and D. Honca).

The National Agricultural Biotechnology Council, annual meeting, May 2001 (D. Birt, C. Scanes, and L. Westgren).

The Center for Judaism and the Environment, and Center for Business Ethics, Jerusalem College for Technology, Israel (A. Wolff, P. Rosenstein, and J. Rose); and "Symposium 2001: Plant Biotechnology, Its Benefits Versus Its Risks", Tel Aviv University, Israel, May 2001 (B. Epel and R. Beachy).

This essay originally appeared as "Ethics and Genetically Modified Foods," SCOPE Research Group (U. California-Berkeley, U. Washington, and American Association for Advancement of Science, July 2001)

In describing this method, I have drawn on an ethics assessment tool devised by Courtney Campbell, Philosophy Department, Oregon State University, and presented at the Oregon State University Bioethics Institute in Corvallis, OR, Summer 1998

References

Comstock G (1988) The case against bGH. Agric Hum Values 5:36–52. The other essays are reprinted in (2) chapt. 1–4
Comstock G (2000) Vexing nature? On the ethical case against agricultural biotechnology. Kluwer Academic Publishers, Boston
Gressel J (1999) Observation at the Annual Meeting of the Weed Science Society of America, Chicago, 10 Feb 1998; Ryan A, et al. Genetically modified crops: The ethical and social issues. The Royal Society, London, section 1.38
Fox J, Hayes D, Shogren J (2002 Jan) Consumer preferences for food irradiation: How favorable and unfavorable descriptions affect preferences for irradiated pork in experimental auctions. J Risk Uncertain 24(1):75–95
Kass L (1988) Toward a more natural science: Biology and human affairs. Free Press, New York; Kass L (1998) Beyond biology: Will advances in genetic technology threaten to dehumanize us all? The New York Times on the Web, 23 August 1998, online at http://www.nytimes.com/books/98/08/23/reviews/980823.23kassct.html
McNeill W (1989) Gains and losses: An historical perspective on farming. The 1989 Iowa Humanities Lecture. National Endowment for the Humanities and Iowa Humanities Board, Iowa City, IA, p. 5
Midgley M (2000) Biotechnology and monstrosity: Why we should pay attention to the 'yuk factor'. Hastings Center Rep 30(5):7–15
Nelkin D, Lindee MS (1995) The DNA mystique: The gene as cultural icon. Freeman, New York

Chapter 5
Responsible Agro-Food Biotechnology: Precaution as Public Reflexivity and Ongoing Engagement in the Service of Sustainable Development

Marian Deblonde

Abstract In Europe, agro-food biotechnologies arouse a lot of controversy. The scope of issues debated during the past four decades of their development has been extending rather than shrinking. And it looks as if many of these issues have been transferred to other new and emergent technologies. This paper considers the adequacy of Europe's regulatory reaction – in the way it interprets and uses the precautionary principle – to respond to these issues. It argues that a fundamental re-interpretation of this principle is needed. It should be re-linked to the guiding idea of sustainable development. This re-linking implies a collective engagement, construction of projections for the future, and a continuous learning process of responsible acting.

Scientific and technological developments resulting in applications of GMOs in agricultural and food practices can be expected to induce, time and again, societal controversies. Attempts have been made to respond to these controversies by a variety of measures in the context of regulation and communication. These attempts do not suffice: they do not succeed in integrating societal issues in adequate ways. What does the concept "societal issues" consist of? What does "adequate integration" mean? In this article, we argue that an adequate integration of societal issues presupposes a different interpretation of the precautionary principle than the one prevailing in European regulation. Precaution should be understood as a necessary attitude to contribute to the guiding idea of sustainable development. What does such a precautionary attitude imply? And how does this interpretation of precaution relate to the European regulatory interpretation?

M. Deblonde (✉)
Institute Society and Technology – Flemish Parliament, Brussels, Belgium
e-mail: marian.deblonde@vlaamsparlement.be

5.1 The Public Debate on Agro-Food Biotechnology

5.1.1 A Short Historical Reconstruction

Devos et al. describe, in a brief historical reconstruction, how public debates regarding agricultural biotechnologies gradually extend the scope of actors involved and issues addressed. Their historical reconstruction illustrates, at the same time that the issues arising relate to the changing ways in which agro-biotechnological research is embedded in societal contexts. Agro-biotechnology research evolves from laboratory science to commercially interesting research toward spearheading national economic and innovation policies. In this section, we will summarize the described evolutions in public debates with regard to GM (genetically modified) crops.

Initially, in the early 1970s, scientists themselves addressed the safety of laboratory workers performing r-DNA research and of the community living in the close neighborhood of molecular biology laboratories. In the USA, a voluntary moratorium on certain types of experiments was even proposed as long as the hazards could not be properly evaluated or prevented. This call for a moratorium induced the first public debates on the r-DNA technique and its applications.

From the beginning, the community of scientists working in the domain of biotechnology was not very eager to share its safety considerations with the public. They feared that involvement of members of the wider public would block scientific research. Nevertheless, the Asilomar II conference (February 1975), intended to discuss appropriate and concrete ways to deal with safety issues of the r-DNA technique, was attended by lawyers, members of the press and government officials as well as scientists. Inclusion of these new actors was a reaction to criticisms on the exclusion of the public. The Asilomar II recommendations led to "Guidelines for Research Involving r-DNA Molecules." These guidelines allowed for the abandonment of the voluntary moratorium and helped to convince US Congress that legislative restrictions were not needed. At the same time, they prohibited large-scale field-testing.

From 1976 onwards, the r-DNA domain evolved from predominantly academic and fundamental research mainly financed by government to applied techno-science largely co-financed by private capital. In this period, patenting of inventions related to the r-DNA technique became an issue. With the adoption of the Bayh-Dole Act (1980), universities, companies and nonprofit organizations obtained the right to hold patents on innovations arising from federally funded research. This augmented the commercial attractiveness of biotechnology. Moreover and in order to maintain leadership in biotechnology, the use of transgenic organisms started to be deregulated. The safety issue was downplayed in favor of industrial expansion.

From 1981 onwards, biotechnology became an important vehicle for economic competition on a national and global scale. Commercial interest in biotechnology extended: not only pharmaceutical multinationals, but also chemical-seed, oil, and agro-food multinationals invested – via the acquisition of biotechnology companies or the funding of fundamental research at universities – in R&D activities in this domain. Various governments established programs to stimulate the industrial development of biotechnology.

Together with a growing commercial exploitation of the technological possibilities of genetic engineering a wider variety of social issues, next to the safety of laboratory workers and of their neighborhoods, arose in public debates. The first deliberate releases of transgenic organisms into the environment in the 1980s generated more focused discussions regarding potential environmental risks. Conflicting positions surfaced with regard to the kinds of harm to take into consideration, the appropriate baseline of comparison to assess the severity and acceptability of risk, the reliability of scientific evidence, and suitable means to manage risks. The fact that knowledge became a commodity for private profit raised a range of questions: about the ethical accountability of patenting natural parts of life forms, about the potential adverse impacts on the accessibility of new developments for scientists and consumers, about the hybridization between fundamental and applied research, about the confidentiality of scientific results, about the moral positions and values of science and researchers, about the credibility and autonomy of scientists, and about the role of government. Sustainability and globalization aspects also entered the debate. The relation between intensive farming – of which GM agro-technology is still a further step – and biodiversity was, for instance, questioned and, therefore, protection or promotion of organic and integrated farming was called for. Moreover, increased alignment of universities with the private sector, a general decrease in crop diversity, and the dominance of the private sector in research, development and commercialization of transgenic crops was considered to cause that the worldwide growing area of GM crops only covered a few crops with a limited variety of engineered traits. Public debate also focused on food quality issues such as the maintenance of farmers' craftsman expertise, of cultural identity and diversity. Concentration of biotechnology companies, undermining of less-intensive farming methods and of high-quality food products, greater farmer dependence on biotechnology companies, and limited opportunities for academic and governmental research institutions and small biotechnology companies – compared to multinational companies – to commercialize transgenic crops are some of the socio-economic issues being discussed.

In the first decade of the twenty-first century, regional and local governments and municipalities and farmers in the EU forged coalitions and installed a network of regions free from GM crops. By this action, they succeeded in defending a model of sustainable rural development against the dominant agro-industrial model. This action urged the EU and national and regional authorities to develop legal co-existence frames. This appears to be an extremely difficult task, as champions and adversaries continue to debate the feasibility of co-existence. Devos et al. conclude that "previously unsolved conflicts over GM crops condensed onto co-existence, which became another arena for contentious values and ideals."

To conclude, in the public debate on agricultural biotechnologies a wide variety of issues matter: next to safety issues, it regards the usefulness of GM crops and the validity of the motives underlying their development, socio-economic impacts of their commercialization, freedom of choice, unnaturalness of genetic modification, respect for nature, irreversibility of adverse effects, democracy, disparities between the industrialized world and the third world, uncertainties with regard to possible risks, fallibility of experts, sustainability effects of agricultural models. The short historical overview illustrates, moreover, that at least some of the issues in the public debate result from specific characteristics in the societal context itself.

5.1.2 An Inadequate European Regulatory Response

In response to societal concerns regarding the harmfulness and scientific uncertainties surrounding GM agro-food products, the EU regulatory framework was gradually restyled. GM agro-food products were subjected to a risk analysis prior to their use. The precautionary principle was explicitly adopted in legislation regarding authorization procedures for deliberate releases of GM agro-food products. The European Food Safety Authority (EFSA) was created to provide "independent scientific advice and clear communication on existing and emerging risks" (www.efsa.europa.eu). Products consisting of, containing or being produced from a genetically modified organism (GMO) were charged to be labeled thus; the traceability of food chains was imposed; a legal framework for co-existence started to be commissioned; and measures to inform the public were installed. Other than risk issues – general ethical issues that were explicitly labeled thus – were channeled to a European ethics committee (European Group on Ethics in Science and New Technologies, http://ec.europa.eu/european_group_ethics/) or to National Ethics Committees. Many ethical issues remained, however, unanswered on a public level and were banished to the private sphere, where individuals can only decide as consumers whether or not to buy particular GM agro-food products.

The restyling of the EU regulatory framework does, until now, not suffice to allay the European public's reservations towards agro-food biotechnology. The Eurobarometer study on biotechnology concludes that "one area where public trust is still lacking is green biotechnology which includes using genetically modified organisms (GMOs) in foodstuffs" (http://ec.europa.eu/research/biosociety/public_understanding/eurobarometer_en.htm, consulted February 1, 2008). A majority of Europeans (58%) believe that the development of GM foods should not be encouraged. GM food is seen by them as not being useful, as morally unacceptable and as a risk for society. The authors of this Eurobarometer report argue that, to build further confidence in science policy, "it would seem prudent to ensure that moral and ethical considerations and the public voice(s) are seen to inform discussions and decisions."

5.2 Precaution in European Regulatory Texts and Practices

5.2.1 **European Regulation...**

In 2002, the precautionary principle was adopted in the EU regulatory framework as one of the ways to respond to societal considerations concerning agro-food biotechnologies. In the European regulatory context, the Communication of the European

Commission on the Precautionary Principle is an important policy document. This Communication places the precautionary principle within the existing framework of risk analysis. "Application of the precautionary principle is part of risk management, where scientific uncertainty precludes a full assessment of the risk and when decision makers consider that the chosen level of environmental protection of human, animal and plant health may be in jeopardy" (COM 2000 1).

The Communication starts from the assumption that the precaution principle is, initially, a management principle that is triggered when potentially dangerous impacts of a phenomenon, product, or process are stated and a scientific evaluation cannot determine the risks with sufficient certainty. The latter condition implies that a scientific evaluation that is as complete as possible and that explains the degrees of uncertainty connected to it is a precondition to take precautionary measures. The Communication reminds us that the risk perception inducing a scientific evaluation is rather of a practical than a theoretical kind. It stresses, further, that a precautionary decision is in the last resort a political decision and that its correctness depends on the societal acceptability of the risks society will have to bear.

The guidelines for application that are proposed in the Communication on precaution are proportionality, nondiscrimination, coherence, considering costs and benefits of both acting and nonacting, considering scientific developments. These guidelines are comparable to the ones that should be applied when other – rather preventative than precautionary – management measures are to be taken.

The interpretation of the precautionary principle as suggested by the European Commission is reflected in the European Communities' Deliberate Release Directives 90/220 and 2001/18. These documents were designed to manage scientific and political uncertainty about hazards of genetically modified organisms. These documents are implicitly precautionary as far as they regulate a priori entire categories of products for which there was not prior evidence of harm. They are explicitly precautionary as far as they declare the precautionary principle as their first priority. Precaution, then, means that intended releases should be assessed case-by-case and step-by-step and that this assessment should be based on expert advice concerning the biosafety, i.e., safety for human health and for the environment (including biodiversity) of the release at stake. Directive 2001/18/EC provides, further, general guidelines concerning the ethical and social aspects of deliberate releases. The European Group on Ethics in Science and New Technologies may be consulted in order to obtain advice on ethical issues of a general nature. Member states of the EU retain, moreover, a competence of their own as regards ethical issues. The latter statement remains, however, to a certain extent a dead letter, since concrete substantive and/or procedural recommendations to integrate social and ethical issues in GMO-policies are lacking.

European regulation on GM crops, thus, focuses on biosafety. The fact that safety assessments should be of a scientific nature reveals, at the same time, other goals. A scientific approach is deemed necessary to contribute to an objective and harmonious treatment of the dossiers. Objective and harmonious procedures should support and stimulate scientific research and innovation, avoid unequal conditions of competition, eliminate impediments – between and within EU countries – while developing and bringing onto the market products containing GMOs. They should, moreover, provide GMO firms with a transparent legal and administrative frame.

5.2.2 And Hazy Compliance

The precautionary principle has a European history of about four decades now. During this period, many authors complained about its ambiguous and vague meaning. According to Löfstedt, one can question whether EU rulings actually abide by the EU's Communication. Both the EU Chemical White Paper and the European Commission Consultation Document on Chemical Regulation call for substances that are persistent, bio-accumulative or known endocrine disrupters to be subject to authorization, in effect leading to a general ban on substances deemed of very high concern. This illustrates that Europe deems the use of the precautionary principle justified even without backing from scientific committees. Similarly, two important legal rulings by the Court of First Instance reaffirmed that precaution should not always be interpreted as part of a risk assessment. These cases arose from a 1999 EU regulation banning antibiotic additives in animal feed on the basis that bacterial resistance to antibiotics might be transferred to humans (though there was no reputable scientific evidence that there was such a transfer).

Comparable discrepancies between regulatory documents and practices are obvious with regard to authorization procedures for GMO releases. Social conflict within many European member states has been preventing a straightforward application of European GMO directives. Some national Biosafety Commissions shifted their regulatory practices to a certain extent in response to public opinion. In Spain, for instance, the National Biosafety Commission (NBC) uses input from public debate to assess certain public concerns even if it does not form part of its own formal risk assessment protocol. The NBC's formal risk assessment protocol requires the evaluation of the pathogenicity, genetic stability, dissemination and survival of GMOs, effects on other organisms and gene transfer. Informally, however, it evaluates a number of additional health and environmental issues raised by non-governmental organizations in the public debate: for instance, the use of marker genes resistant to antibiotics, the development of resistance in pests to insect resistant crops as well as the effects of these crops on populations of beneficial (nontarget) insects, and issues related to herbicide use. Moreover, the NBC applied, in a few isolated cases, an implicit technology assessment. Although the scope of the European GMO directives is limited to evaluating health and environmental effects and deciding about the acceptability of those effects, the NBC assessed, for instance, the overall impact on herbicide use of herbicide-resistant crops. It, thus, assessed the expected benefits and compared them to possible risks or costs. Further-reaching demands, like the evaluation of the technology's socio-economic effects, its impact on traditional agriculture or its effects on the North–South relationship, are, however, still excluded by the Commission from the evaluations.

In Belgium, the Biosafety Advisory Council is responsible for assessing authorization requests for deliberate releases of GMOs (www.biosafety-council.be). The Section on Biosafety and Biotechnology (SBB) of the (federal) Institute of Public Health acts as the secretariat of the Council. The SBB watches punctually that the discussions taking place in the scientific committees advising the Council conform to European and Belgian regulations. With regard to the dossiers discussed in 2002, for instance, the chairperson cut short discussions concerning the scientific and societal use of the intended releases and their possible damage for organic agriculture. The Council gave a positive advice to all of the six new authorization requests for field trials it received in 2002. The Competent Authority, however, admitted only four of them, thereby responding to concerns of the wider Belgian public.

In 2003 and 2004, hardly any authorization requests for deliberate releases were submitted in Belgium. In 2005 and 2006, no authorization requests for field trials with GM corps were submitted. However, respectively 20 and 8 authorization requests for the commercialization of GMOs were submitted under Regulation (EC) 1829/2003. [This regulation allows, indeed, gaining at once permission for the import, processing, cultivation-domains regulated by Directive 2001/18/EG, and use as food or feed-domains only regulated by Regulation (EC) 1829/2003. This explains why most authorization requests for the commercialization of GMOs are submitted under this Regulation. Only authorization requests that do not relate to food or feed – as, for instance, genetically modified flowers – can presently be submitted under Directive 2001/18/EG (see the annual reports 2007 and 2006 on www.biosafety-council.be)].

Despite lasting public skepticism towards agro-food biotechnology, continuing disputes between the EU and the USA with regard to the World Trade Organization (WTO) ruling for biotech crops, and complaints about the "capricious and arbitrary" use of the precautionary principle in European Union courts, both public and private investments in research, development, commercialization and cultivation of GM crops are rising worldwide. Monsanto, for instance, seems to regain firm confidence in the future of biotech crops. Since the start of the cultivation of GM crops in 1996, the global area of biotech crops continues to climb with a sustained growth rate of 13%. In 2006 the global area of biotech crops reached 102 million hectares (http://www.isaaa.org/resources/publications/briefs/35/executivesummary/default.html). The amount of genetically modified food products to be found on the supermarket shelves is expected to increase in the years to come. And new research domains, such as nano-biotechnology and synthetic biology, that have a much larger potential for redefining and (re)constructing food and life forms – and, consequently, human worlds – than actual techniques of genetic modification, are starting to be explored. At present (October 2008), it looks as if the active involvement of the wider public with actual R&D activities in the agro-food domain is at a low level and as if the room for reflection upon social issues and for their integration into innovation and authorization policies is even more futile than before.

5.3 Precaution and Sustainable Development

5.3.1 Precaution Lost Its Orientation

Many texts have been written to make the meaning of the precautionary principle more concrete and, consequently, to make the concept operational in a more unequivocal way. In most of these texts the link between precaution and sustainable development remains unclear or is even absent. However, the frequently quoted Bergen definition of the precautionary principle states: "In order to achieve sustainable development, policies must be based on the Precautionary Principle. Environmental measures must anticipate, prevent and attack the causes of environmental degradation. Where there are threats of serious or irreversible damage, lack of full scientific certainty should not be used as a reason for postponing measures to prevent environmental degradation" (Declaration of the United Nations Economic Commission for Europe, May 1990, as cited in Sandin 1999). This declaration suggests that there is an obvious link between sustainable development – or at least its ecological dimension – and precaution, namely that an ecologically sustainable environment cannot be reached without a precautionary attitude.

This link between (even only the ecological dimension of) sustainable development and precaution has been broken in European regulation. According to the Communication, the precautionary principle should aim at a balance between the freedoms and rights of persons, enterprises, and organizations on the one hand and the necessity to limit the risks of negative impacts on the environment and the health of humans, animals, and plants on the other. As Jensen argues, this balancing exercise fits within an ethics of political liberalism. According to this ethics, public authorities should protect the rights and freedoms of individuals and of legal entities such as enterprises and organizations. The only reason for which persons may be restricted in their actions by the use of coercion is to prevent unacceptable harm to entities – in this case humans, animals, and plants – worthy of protection. The task of public authorities consists, within this liberal tradition, of creating a general legal and institutional frame within which economic actors are allowed to act and trade. Within this general liberal framework, restriction on free trade can only be justified in order to prevent harm to third parties.

This liberal interpretation of precaution implies a harm-, rather than goal-oriented approach. The precautionary principle is only applied when an assumption of potential risk exists. The main question is whether it is sufficiently plausible that a particular application will cause so much harm – independent from possible advantages – that precautionary measures should be taken. It is the degree of uncertainty of plausible and sufficient harm (sufficient evidence of sufficient harm) that triggers the idea of precaution.

This harm-oriented approach contends with a serious problem. Detached from explicit goals, it is not clear what counts as "unacceptable harm." The identification of potentially harmful effects involves unavoidably a number of value judgments. Judgments are, for instance, made about which kinds of harm to assess and which to ignore, about what baseline to use for assessing harm, and even about what counts as a harmful impact. Should one, for example, compare the impact of GM crops with the impact of conventional agricultural practices or with the impact of organic agriculture (as Austria would like it)? Does one only count direct harm or also indirect harm (such as the impact of changes in herbicide use as the result of the introduction of herbicide-resistant crops)? And does only harm to nonagricultural land count as environmental harm or also (economic) harm to agricultural land?

In the present European regulatory context, particular technological applications form the starting point for implementing a precautionary attitude. GM crops, for instance, are a type of technological application for which a precautionary attitude is deemed fit. Biotech industries object to this pre-selection. They argue that new crop varieties should be evaluated according to their characteristics (as is the case in Canada), not according to the kind of technology with which they were created. They oppose, hence, the a priori precautionary approach taken with regard to GM crops (and, thus, the necessity of a separate authorization procedure), while new crop varieties with comparable characteristics that are, however, created with more conventional techniques are seemingly not to blame. In line with this course of reasoning, they stress the importance of the principle of familiarity in order to evaluate potential risks of GM crops. (In the European directives, the principle of familiarity is mentioned as the second priority, next to the principle of precaution, to assess authorization requests.)

From a liberal interpretation of precaution, the restriction of an a priori precautionary approach to particular types of technological applications is, indeed, hard to justify. In case public authorities do not deem it necessary to assess more conventional agricultural techniques, one cannot easily provide counterarguments to the conviction of biotech industries that a separate assessment procedure of GM crops is discriminating. Things change, however, when one considers a goal-oriented interpretation of precaution. In case one takes predefined goals, rather than particular technological applications, as the starting point and develops and selects technological applications as a function of these goals, precaution should apply to the whole variety of technological applications that are selected (rather than to none of them, as biotech industries suggest).

5.3.2 Re-linking Precaution to Sustainable Development

If, as the previous line of reasoning suggests, a goal-oriented interpretation of the precautionary principle is needed in order to allow for coherent and nondiscriminatory authorization policies, what is, then, this goal? In order to work out an answer to this question, we reflect a while on the wider societal context within which the principle surfaced as a policy principle.

According to Boehmer-Christiansen, the German concept of *Vorsorge* integrates three meanings: caring for, worrying about, and obtaining provisions. In order to help us clarify the concrete meaning of each of these verbs, we could start with a reflection on the following questions. Should the choice of newly developed technological applications be subordinate to predefined goals, or should the choice of the goals to be realized be subordinate to already developed technological applications? Should the freedom of producers and consumers be subordinate to commonly shared visions on humane conditions of existence both for present and future generations, or should present and future conditions of existence be subordinate to the maximum freedom – even in the sense of just distribution between present and future generations – of producers and consumers? Answers to these questions cannot be of a black-or-white type. For, to start with, it does not make so much sense to define abstract societal goals that are hardly connected to (presumed) technological possibilities on the one hand. And new technological applications are not developed in complete independence from already existing societal goals on the other hand. There always exists some reciprocity between (dominant) societal goals and (dominant) technological developments. Something comparable holds for the relationship between consumers' and producers' freedom and visions on humane conditions of existence. The kinds of freedoms that get legally protected and (dominant) interpretations of humane conditions of existence are influencing each other mutually. Both questions are, though, relevant because they express a need to make these reciprocities and mutual influences more explicit and to create, as a consequence, possibilities to adjust them. This need is illustrated by the many controversies – both scientific and societal – that exist concerning societal introductions of new technological applications. These controversies testify to a need to reflect on and to recalibrate the choices and values embedded in these technologies on the one hand and in the visions and concerns of the wider public on the other.

What makes this recalibration exercise pressing? Following Jonas, this is because of the technological capabilities present in our modern Western societies. Contrary to pre-modern times, human technological and scientific powers are such that the natural conditions of human existence can be altered, either gradually or suddenly. Nature – both the nature of living beings and their environment – proves to be susceptible to the interferences of modern technologies. Our climate shows changes due to the use of greenhouse gases and fossil fuels. Environmental pollution caused by industrial production and consumption patterns has carcinogenic effects on human health. Agro-industrial practices threaten existing biodiversity. The use (or misuse) of nuclear energy – either for applications during peace or war times – threaten the genetic codes of living beings. We are capable of radical changes, both in the short and the very long run and both locally and globally. And because we are capable, we are responsible. It is this dimension of modern, industrial technological powers and the possible threats they entail for the natural conditions of human existence that make a reflection on and recalibration of the values and choices inherent in technological applications, on the one hand, and human concerns and expectations, on the other, pressing.

From now on, according to Jonas, acting technologically is acting in an ethically sensitive way. Our responsibility for humans urges us to take responsibility for nature, since humane conditions of existence depend on it. However, this responsibility does not any longer remain restricted to what happens here and now, because the – often unpredictable, cumulative, and irreversible – effects of our technological actions extend widely in time and space. Our responsibility is, therefore, a responsibility for the future arising from our collective technological acting.

The sense of this responsibility is to safeguard the humanity of the conditions of existence of both present and future generations. Maintaining humane conditions of existence (in the near and far future) has become the decision criterion for our technological acting. According to Jonas, this responsibility is total, continuous, and future-oriented. We are *totally responsible*: not only for the material needs, but also for everything that enables human beings to develop in a humane way (knowledge, social and moral skills, practical and cultural skills, societal structures, and so on). Human beings are, indeed, in the first place responsible for the ability of other human beings to bear, in due course, their own responsibility and to give shape to their own humaneness. We are *continuously responsible*: our responsibility never stops. Our responsibility has a historical dimension: it relates the past with the present and the future. It recognizes what has been handed down – both positive and negative deeds, both failed and performed actions – and asks itself how to integrate them in the future of the people who will live in the future. It regards the tradition of a collective humane identity. We are in the first place *responsible for the future*. A paradox is at the back of this. We are responsible for a future that escapes the effectiveness of our actions. For the results of our actions are unpredictable. They escape our control. We cannot be responsible for the concrete deeds of future generations. For this is precisely the aim of our responsibility towards the future: we should not so much fix the future, but create the conditions so that those living in the future will be able to create their own concrete lives and bear responsibility for their own future.

To summarize, we could apply a more recent terminology to indicate the sense of our responsibility of the future: this sense is presently called sustainable development. The responsibility emerging from our technological possibilities regards the realization of sustainable conditions of existence. This responsibility does not remain restricted to the natural/ecological dimensions of humane conditions of existence. It regards no less their social and economic dimensions.

5.3.3 Dimensions of a Precautionary Attitude

According to Dupuy and Grinbaum, this responsibility for the future asks for a radically new ethics, an "Ethics for the Future" "meaning not a future ethics, but an ethics *for* the future, for the sake of the future, i.e. the future must become the major object of our concern." An ethics for the future urges us to try to anticipate the possible consequences of our technological choices, to assess them and to ground our choices on this assessment. However, since our technological power for acting largely goes beyond our scientific power for predicting the effects thereof, our obligation to anticipate the future is an impossible ambition. If our scientific predictive power were reaching as far as the causal scope of our technological acting, we would not need to be precautionary. This is, however, not possible. And this is, according to Jonas (and Arendt), not a completely new phenomenon. Scientific predicting power is by definition not adequate with respect to political acting: the spontaneity typical for political acting makes it "irrational" from a scientific perspective. The possibly huge scope, both in time and space, and irreversibility of the effects of new technologies add, however, a new dimension. We are aware that present-day technological developments can possibly disturb the humane conditions of existence thoroughly and without leaving us the opportunity to regain control. It is this dimension that urges us to take personal responsibility for the future, despite our unavoidable lack of knowledge of the future. This handicap does not allow us to evade our responsibility. "The precautionary principle implies the need, as a matter of cultural change, for society's institutions to enlarge existing notions of ethical responsibility to encompass these unknowns, which are predictable in principle even though not in specifics." Or, in the words of Dupuy and Grinbaum, we cannot but violate one of the foundations of ethics, namely the principle that *ought* implies *can*.

Precaution could be an answer to this ethical perplexity, on condition that it is differently interpreted than is presently the case in European regulatory contexts. If sustainability is the *sense* of our responsibility for the future, then precaution is the *attitude* that is necessary in order to realize it. What does such a precautionary attitude stand for? A precautionary attitude in the service of sustainable development urges us, first, to replace a harm-oriented approach by a goal-oriented approach. Second, it implies collective engagement, both of public authorities, stakeholders and the wider public. Third, it urges us to base our present choices on projections of the future. And, finally, it presupposes a continuous process.

5.3.3.1 A Goal-Oriented Approach

A goal-oriented approach implies that it is particular sustainability objectives, rather than particular technologies that trigger assessment procedures. It implies, moreover, that these assessment procedures are of an integral kind. This means (a) that not only environmental and health issues of technological applications are evaluated, but also social, economic, and ethical ones, (b) that both advantageous and disadvantageous issues are assessed, so that a weighing can take place, and (c) that advantages and disadvantages of various possible solutions – both technological and other ones – are compared. Integral assessment presupposes, hence, that research and innovation policy stimulates technological diversity against the background of publicly defined sustainability goals (and diversity of the social practices to which the respective technologies belong).

At first sight the European communication seems not to contradict this idea of an integral evaluation. Three considerations, described in the communication, are important to mention in this respect. First, regarding the criterion of proportionality – measures should bear a proper proportion to the intended protection level – the communication suggests that possibilities to replace intended products or processes by less dangerous ones should be taken into consideration. Second, regarding the criterion of weighing costs and benefits it states, to begin with, that a balancing of costs and benefits should not consider only economic data, though an economic cost-benefit analysis should be part of the investigation whenever such an analysis is feasible and desirable. And, what is more important, it states further that an analysis of the effectiveness of several possible options and of their acceptability for the public should be examined, since it is conceivable that society is prepared to pay a higher price in order to guarantee an interest deemed primordial. Third, regarding the criterion of considering scientific developments, the communication puts forward that one should look for better methods and instruments for risk-evaluation that include all relevant factors such as, for instance, social-economic information and technological perspectives.

The problem is, however, that in the European regulatory context these considerations only achieve meaning in the phase of deciding on appropriate precautionary measures, i.e., after a previous risk assessment with regard to a particular technological application has taken place and the conclusion fell that sufficient evidence of sufficient harm exists. In such a harm-oriented approach, the search for alternative technological applications cannot begin before the plausibility of sufficient harm is established. This implies that alternative technologies are not considered when the harmfulness of a specific technological activity – either or not with the help of small restrictions or adjustments – remains below a certain level that is deemed acceptable. The preceding risk assessments to be made, according to directive 2001/18, are limited to environmental impacts, "thereby missing the economic and social dimensions of sustainable development." A harm-oriented approach implies, moreover, that, in case consideration of possible alternatives does happen, it begins late and in an unsystematic way. One tends, first, to investigate only those alternatives that bear close resemblance to the initial technology. Second, experts of advisory bodies are often not in a position to compare alternatives: they have to advise on products that are already on the market or that will be introduced very soon, or possible alternatives (and their potential risks and benefits) are hardly known. The fact that many R&D activities taking place within private enterprises are subject to secrecy clauses does not stimulate a scientific comparison of possible alternatives. Third, in some cases, European regulation even forbids a comparison of alternatives. It is, for instance, legally not allowed to prohibit a pesticide because it is less good than an existing alternative.

5.3.3.2 Collective Engagement

Several considerations justify the engagement of the whole society as a dimension of a precautionary attitude. To start with, the concept of sustainable development as the sense of a precautionary attitude should be reflected in a public ethic. Sustainable development is an ethical guideline belonging to the public rather than the private sphere, because it is the result of our collective acting that affects the conditions of human existence. It applies, hence, to the institutional organizations defining individuals' scope of choice for acting. This interpretation sounds like an anachronism. We are not used to thinking of public authorities as defending, explicitly, a common ethic from a perspective that transcends the present, that links the future to our present and our past. We are rather used to interpreting public authorities as utilitarian institutions that should defend the actual safety of their citizens, but without interfering with the actual (predefined) freedoms of producers and consumers.

Second, sustainable development is an ethical guiding line that should be collectively defined. In order to make the concept of sustainable development concrete with regard to technological developments, public debates are needed. Individual scientists, engineers, and experts cannot take responsibility for their discoveries and engineering designs because these discoveries and designs get transplanted into the subsystems of economy, politics, and law and, hence, transformed according to the specific logics of these subsystems. These system logics are not traceable to the intentions of particular individuals, nor are the possible, but unintended and often not assessable consequences of the transplanted and transformed scientific and technological applications. Therefore, Mitcham and von Schomberg argue, all citizens should respond personally. Personal responsiveness means that individual participation in public debates is the default position: persons must give reasons for being excused from such a duty. Public deliberation serves the function of presenting different relevant issues to the more or less autonomous systems and subsystems of society, i.e., to politics, law, science, and so on. Appropriate exchanges between the various subsystems and the wider public are needed. Representatives of these subsystems need to respond to publicly identified and articulated issues. Conversely, they are drivers for new debates when they publicize particular aspects of an issue that cannot be fruitfully resolved within the limits of the typical specialized discourse of the subsystem they belong to.

Third, engagement of wider society is needed to influence the definition and selection of socio-technological trajectories that can contribute to the realization of sustainable development. In the past decades, the conviction that techno-scientific developments do not happen autonomously, but result from many choices made by people within various human-made institutions gradually gained ground. Science, technology and society co-evolve in many complex ways. Techno-scientific developments are open to individual creativity, collective ingenuity, economic priorities, cultural values, institutional interests, stakeholder negotiation, and the exercise of power. They are inherently open to human agency and to deliberate social choice. The big challenge is to make them open to the interests of sustainable development – as collectively defined – rather than to a narrow set of incumbent interests. In order to realize that, collective involvement is needed to counterbalance unequal distributions of power. Actual processes of corporate concentration, institutional harmonization, and economic globalization render the governance of science and technology ever more obscure and inaccessible. Therefore, empowerment of the wider public in order to realize wider social agency in institutional and techno-scientific choices is highly needed. Moreover, collective involvement – from the very beginning of innovation trajectories – is needed because successful implementation of (sustainable) socio-technological trajectories depends on the reflexivity, creativity and commitment to new initiatives of various stakeholders, be it regulators, producers, consumers, or users.

5.3.3.3 Projections of the Future

How can we base our present decisions on projections of the future, given the many societal and scientific uncertainties surrounding new technologies? Dupuy and Grinbaum argue that we should build both self-fulfilling and self-denying prophecies. They argue for "ongoing normative assessment." This method is, on the one hand, "a matter of obtaining through research, public deliberation, and all other means, an image of the future sufficiently optimistic to be desirable and sufficiently credible to trigger the actions that will bring about its own realization." It is, on the other hand, a matter of "enlightened doom saying," i.e., of obtaining "an image of the future sufficiently catastrophic to be repulsive and sufficiently credible to trigger the actions that would block its realization, barring an accident." This way of relating to the future presupposes a particular interpretation of time. Time should not be understood as something that can be predicted, nor as something that occurs depending on decisions taken at particular moments. Time should be understood as a closed loop: past and future reciprocally determine each other. In "projected time" the fact is taken seriously that anticipations of the future influence the past, while in "occurring time," causal influence is deemed to flow unidirectionally from the past to the future.

Since the future depends on the way it is anticipated, every determination of the future must take into account the language that is being used to describe the future (visions, expectations, hopes and fears) and how this language finds its way to public opinion and into public and private innovation strategies. Many futuristic visions or techno-scientific imaginaries have, however, an inherent bias to desirable futures, because they are constructed by the enactors of these new technologies. Enactors have an interest in moral and financial support for their R&D activities. Therefore, they are eager to disseminate positive imaginaries and fear "outsiders" who describe "dark" scenarios. However, as Dupuy and Grinbaum argue, building both self-fulfilling and self-denying prophecies is needed as a way to deal with fundamental and insurmountable uncertainties. Here is another reason to involve the wider public: since the wider public does not reflect from an enactor perspective, but from the perspective of a comparative selector, they feel freer to consider both advantages and disadvantages, attractive and threatening dimensions of new technologies.

It is not enough to construct abstract visions of the future, i.e. techno-scientific imaginaries – or roadmaps in the parlance of technological industry – considered separate from the societal context that will welcome particular technologies and that will embed them in specific ways. Issues accompanying new technological developments are often not inherently linked to the technologies themselves. Whether new technologies will alleviate or exacerbate particular societal goals depends on the particular technological applications that are selected; on how they are implemented, disseminated, and situated (and on who or what factors determine these); on who controls them; and on what sorts of oversight and regulations pertain to them (and how effectively these are enforced). Technologies are not separable from their social context. In order to identify and respond to the social and ethical issues associated with new and emergent technologies, it is no use to contemplate these developments as such, in the abstract, without situating them in their social context. Issues rather emerge from new technologies' interactions with problematic features of the social, ethical, cultural, and institutional contexts into which they are emerging.

This implies that a responsible governance of R&D activities cannot remain restricted to solving problems of technology-design or risk management and requires much more than maintaining open lines of communication and engaging in public outreach. To fully address the social, economic and ecological issues associated with new technologies requires addressing socially, economically and ecologically problematic features of the social contexts into which new technologies are emerging (Sandler 2007). The social and ethical challenges of new technologies can be fully identified only if both the characteristic features of these technologies and the social contexts into which they are emerging are considered. The potential of technologies can, indeed, turn out very differently depending on the societal context in which they develop and get embedded. When making projections, the construction of a fictive, but coherent societal context is apparently needed to let a more complete picture of possible issues relating to particular technologies become manifest. The credibility of these projections depends on the extent that they are based on an analysis of (both problematic and promising) features of the actual societal context.

Once the issues are identified, they can only be addressed by remedying the problematic features of the social contexts, which cannot be accomplished by technology design or risk management alone. This claim enforces the argument that responsible governance of new technologies is a matter of collective co-responsibility (von Schomberg 2007). It is a responsibility that cannot be born by enactors of new technologies alone. Responsible governance needs the commitment of the whole society: in order to identify these problematic features, in order to keep them on the agenda, and in order to take initiatives to change them.

5.3.3.4 A Continuous Learning Process

A goal-oriented approach presupposes (even more than a harm-oriented one) a continuous learning process, for scientists concerned as well as for citizens and policy people. A goal-oriented approach is characterized by the circular temporality of projected time. Influences between the past and the future go both ways, from the system to the observing and acting members of society and from these members to the system. Because of these mutual interactions, all the predictive work cannot be done at one and only one point of time. The circularity of relations requires that predictions be constantly revised, in order to make consistent updates. The implementation of new technologies involves many uncertainties. A continuous monitoring of effects is, consequently, necessary. Some technological alternatives can prove to be more and others less harmful than initially expected. Or it can turn out that one urgently needs to look for new alternatives. Co-operation from all sides of a society and transparent participatory decision-making processes are needed to implement such learning processes.

This idea of a continuous learning process is at odds with the guiding principles of nondiscrimination and coherence, as suggested by the Commission. However, if the precautionary principle is expected to respond to previously defined societal objectives, changing societal concerns and changing scientific information can, indeed, ask for stricter precautionary measures than had been applied in the past in comparable situations.

5.4 Conclusion

Techno-scientific developments in the domain of agro-food technology are evolving rapidly and will possibly have far-reaching transformative powers for the food we will eat in the near future, for agricultural and agro-industrial practices, and for the wider conditions of human existence. Until now, integration of the precautionary principle in European regulatory practices, next to other trials, for instance, to regulate labeling and co-existence, do not seem to suffice to quiet public opinion. In this article, we argue that a more adequate response to public controversies needs a fundamentally different interpretation of the precautionary principle. Precaution should be re-linked to the guiding idea of sustainable development. Precaution is a necessary attitude to allow for sustainable development. This attitude consists of integrated assessments, based on projections of how new technologies will be embedded in societal contexts. These projections – both attractive and repulsive, but in any case credible ones – should function as references for actual decisions. Both these projections and the integrated assessments should be the subject of an ongoing process that engages the whole society. Defining and organizing this process, and distributing and defining the responsibilities of various members of society in this process, may look an enormous task. Compared to the revolutionary powers induced by the fast, mainly economically driven techno-scientific developments, it is, perhaps, the only suitable response.

References

Arcuri A (2007) Reconstructing precaution, deconstructing misconceptions. Ethics Int Aff 21(3):359–379

Bodansky D (1991) Scientific uncertainty and the precautionary principle. Environment 33:43–44

Boehmer-Christiansen S (1994) The precautionary principle in Germany – Enabling government. In: O'Riordan T, Cameron J (eds) Interpreting the precautionary principle. Earthscan, London, pp. 31–60

Calman K, Smith D (2001) Works in theory but not in practice? The role of the precautionary principle in Public Health Policy. Public Admin 79:185–204

Carr S (2002) Ethical and value-based aspects of the European Commission's precautionary principle. J Agric Environ Ethics 15:31–38

COGEM (2003) Naar een integraal ethisch-maatschappelijk toetsingskader voor moderne biotechnologie.

COM (2000) Communication from the Commission on the precautionary principle

Deblonde M, du-Jardin P (2005) Deepening a precautionary European policy. J Agric Environ Ethics 18:319–343

Deblonde M, Van Oudheusden M, Evers J, Goorden L (2008) Co-creating nano-imaginaries: Report of a Delphi-Exercise. Bull Sci Technol Soc 28:372–389

DeKay ML, Small MJ, Fischbeck PS, Farrow RS, Cullen A, Kadane JB, Lave LB, Morgan MG, Takemura K (2002) Risk-based decision analysis in support of precautionary policies. J Risk Res 5:391–417

Devos Y, Maeseele P, Reheul D, Van Speybroeck L, Dewaele D (2007) Ethics in the societal debate on genetically modified organisms: A (re)quest for sense and sensibility. J Agric Environ Ethics 20:33–39

Dovers SR, Handmer JW (1995) Ignorance, the precautionary principle, and sustainability. Ambio 24:92–97

Dupuy J-P, Grinbaum A (2006) Living with uncertainty: Toward an ongoing normative assessment of nanotechnology. In: Schummer J, Baird D (eds) Nanotechnology challenges. Implications for philosophy, ethics, society. World Scientific, Singapore

Gaskell G, Stares S, Allansdottir A, Allum N, Corchero C, Fischler C, Hampel J, Jackson J, Kronberger N, Mejlgaard N, Revuelta G, Schreiner C, Torgersen H, Wagner W (2006) Europeans and biotechnology in 2005: Patterns and trends

Goorden L (2003) Finding a balance between technological innovation and deliberation. Lessons from Belgian Public Forums on biotechnology. In: Annual Meeting of the American Political Science Association. Philadelphia, USA

Haag D, Kaupenjohann M (2001) Parameters, prediction, post-normal science and the precautionary principle – A roadmap for modeling for decision-making. Ecol Model 144:45–60

Harremöes P, Gee D, Macgarvin M, Stirling A, Keys J, Wynne B, Guedes Vaz S (2002) The precautionary principle in the 20th century. Late lessons from early warnings. Earthscan, London

Henry C, Henry M (2003) L'essence du principe de précaution: la science incertaine mais néanmoins fiable.

Jensen KK (2002) The moral foundation of the precautionary principle. J Agric Environ Ethics 15:39–55

Jonas H (1984) The imperative of responsibility. In search of an ethics for the technological age. University of Chicago Press, Chicago

Karlsson M (2003) Biosafety principles for GMOs in the context of sustainable development. Int J Sustain Dev World Ecol 10:15–26

Levidow L (2001) Precautionary uncertainty: Regulating GM crops in Europe. Soc Stud Sci 31:842–874

Löfstedt RE (2004) The swing of the regulatory pendulum in Europe: From precautionary principle to (regulatory) impact analysis. J Risk Uncertain 28:237–260

Löfstedt RE, Fischhoff B, Fischhoff I (2002) Precautionary principles: General definitions and specific applications to genetically modified organisms. J Policy Anal Manage 21:381–407

Marchant GE, Mossman KL (2004) Arbitrary & capricious. The precautionary principle in the European Union Courts. The AEI Press, Washington, DC

Mayer S, Stirling A (2002) Finding a precautionary approach to technological developments. Lessons for the evaluation of GM crops. J Agric Environ Ethics 15:57–71

Mitcham C, von Schomberg R (2000) The ethic of scientists and engineers: From occupational role responsibility to public co-responsibility. In: Kroes P, Meijers A (eds) Research in philosophy and technology. JAI Press, Amsterdam [etc.], pp. 167–189

Morris J (2002) The relationship between risk analysis and the precautionary principle. Toxicology 181–182:127–130

O'Brien M (2003) Science in the service of good: The precautionary principle and positive goals. In: Tickner JA (ed) Precaution, environmental science and preventive public policy. Island Press, Washington/Covelo/London, pp. 279–295

O'Riordan T, Cameron J (1994) Interpreting the precautionary principle. Earthscan, London

O'Riordan T, Jordan A, Cameron J (2001) Reinterpreting the precautionary principle. Cameron May, London

Punie Y, Maghiros I, Delaitre S (2006) Dark scenarios as a constructive tool for future-oriented technology analysis: Safeguards in a world of ambient intelligence (SWAMI), Second International Seville Seminar on Future-Oriented Technology Analysis. Impact of FTA Approaches on Policy and Decision-Making. Seville

Raffensperger C, deFur L (1999) Implementing the precautionary principle: Rigorous science and solid ethics. Hum Ecol Risk Assess 5:933–941

Ricci PF, Rice D, Ziagos J, Cox LAJ (2003) Precaution, uncertainty and causation in environmental decisions. Environ Int 29:1–19

Rip A (2006) Folk theories of nanotechnologists. Sci Cult 15:349–365

Sandin P (1999) Dimensions of the precautionary principle. Hum Ecol Risk Assess 5:889–907

Sandin P, Peterson M, Hansson SO, Ruden C, Juthe A (2002) Five charges against the precautionary principle. J Risk Res 5:287–299

Sandler R (2007) Nanotechnology and social context. Bull Sci Technol Soc 27:446–454

Sarewitz D (2005) This won't hurt a bit: Assessing and governing rapidly advancing technologies in a democracy. In: Rodemeyer M, Sarewitz D, Wilsdon J (eds) The future of technology assessment. Woodrow Wilson International Center for Scholars, Washington, DC

Starr C (2003) The precautionary principle versus risk analysis. Risk Anal 23:1–3

Stirling A (2008) "Opening up" and "Closing down": Power, participation, and pluralism in the social appraisal of technology. Sci Technol Hum Values 33:262–294

Tickner JA (2003) Precaution, environmental science and preventive public policy. Island Press, Washington/Covelo/London

Todt O (2004) Regulating agricultural biotechnology under uncertainty. Safety Sci 42:143–158

Treich N (2000) Décision séquentielle et principe de précatuion. Cahiers d'économie et sociologie rurales 55–56:6–24

Turner D, Hartzell L (2004) The lack of clarity in the precautionary principle. Environ Values 13:449–460

van den Belt H, Gremmen B (2002) Between precautionary principle and 'sound science': Distributing the burdens of proof. J Agric Environ Ethics 15:103–122

von Schomberg R (2007) From the ethics of technology towards an ethics of knowledge policy & knowledge assessment. European Commission, Brussels

Wilsdon J (2005b) Paddling upstream: New currents in European technology assessment. In: Rodemeyer M, Sarewitz D, Wilsdon J (eds) The future of technology assessment. Woodrow Wilson International Center for Scholars, Washington, DC

Chapter 6
Precautionary Approaches to Genetically Modified Organisms and the Need for Biosafety Research

Anne Ingeborg Myhr

Abstract Many interpretations of the precautionary principle do not provide a basis for adequate decision-making in regard to ecological and ethical questions. The article disagrees with the opinion that the precautionary principle is not scientific and gives reasons for an enlargement of scientific methods. The author agitates for an argumentative use of the concept, which has sustainable acting as its fundament.

6.1 Introduction

The emergence of biotechnology into everyday life has transformed much of our world experience: opportunities, choices, risks, and irrevocabilities. In particular, the commercialization of genetically modified (GM) crops has raised a polarized debate concerning risk, social, cultural, and ethical issues. Scientists and representatives of the biotechnology industry have often dominated debates concerning risk issues (Melo-Martin and Meghani 2008). There are many areas concerning use and release of GM crops and genetically modified organisms (GMOs) in general, where there is still insufficient scientific understanding, both with regard to intended and unintended impacts. Scientific data concerning environmental and health effects have been conflicting and highly dependant on one's starting assumptions, background, and values. This has caused disagreements about the significance of scientific evidence in relation to benefits and risks and the appropriateness and necessity of biosafety frameworks and regulation. Especially, the role of the precautionary principle in risk assessment and management processes of GMOs has caused controversy.

The precautionary principle has been accepted by many national governments as a basis for policy-making, and it has become important both in international

A.I. Myhr (✉)
Genøk – Centre for Biosafety, Tromsø, Norway
e-mail: anne.i.myhr@uit.no

environmental law and international treaties. With regard to GMO regulations, the precautionary principle plays an important role in the Cartagena Protocol, an international agreement on the trans-boundary movement of GMOs, and in regulations such as the Norwegian Gene Technology Act of 1993 and the EU directive 2001/18/EC.

Implementing a precautionary approach requires a renewed look at the science underpinning risk assessment and management of GMO use and release. This implies that there is a need to identify practical means for precautionary-oriented research design and to explore the role of precautionary-motivated science in policy and research decisions. There is a need for more comprehensive studies of ecological effects of GMO utilization, for instance with regard to potential unintended effects of GMO release on environmental processes and adverse effects on human and animal welfare (Myhr and Traavik 2003). Experimental testing of carefully elaborated risk hypotheses may result in a solid basis for the avoidance of potentially harmful GMOs. The initiation of such risk-associated research involves some challenges since it questions the traditional conduct of science, i.e. with regard to reliance on methods and the choice of hypothesis. There is a need to achieve wise management of uncertainties with regard to potential adverse effects of GMO release. A more precautionary approach to GMO risk issues implies awareness of the complexity of ecological aspects; it involves broadening the scientific focus, and allows for interdisciplinary approaches. Interpretation of the uncertainties involved is partly a scientific and partly a normative question, hence it is necessary to establish normative baselines concerning uncertainty related to risk and beneficial aspects. Such an approach may reduce the present controversies surrounding GMO use and release, and hopefully be robust and transparent, which may increase trust and legitimacy. These issues will be elaborated upon in the rest of this chapter.

6.2 The Precautionary Principle: Definitions and Applications

The precautionary principle is a normative principle for making practical decisions under conditions of scientific uncertainty. The employment of the precautionary principle entails the identification of risk, scientific uncertainty and ignorance, and it involves transparent and inclusive decision-making processes (Kriebel et al. 2001).

At present the employment of the precautionary principle in risk assessment and management of GMO use and release is subject to heated scientific and public controversies. In the view of the opponents, employment of the precautionary principle may place an additional regulatory burden on GMO utilization, and thereby reduce returns from innovation, limit utilization of GMOs worldwide, and provide disincentives for research. On the other hand, advocates of precautionary principle employment want to enhance competitive safety procedures and to separate trade and environmental interests in decisions. Most criticism of the precautionary principle is based on comparison with qualitative risk assessment techniques (Stirling 2007), and the critique is often based on misinterpretations (Sandin 2004). Implementation of the precautionary principle means that more, not less, scientific research is needed for a comprehensive view of how to achieve environmental and public health protection.

Several versions of the principle, ranging from ecocentric to anthropocentric, and from risk-adverse to risk-taking positions, have been put forward. Sandin (2004) makes the distinction between argumentative and prescriptive versions of the principle. Prescriptive versions are passive in nature and grant regulators permission to take action to prevent harm, but lacks requirement for action.

Examples of treaties that employ prescriptive versions are for instance the Rio Declaration and the Cartagena Protocol on Biosafety. In the Rio Declaration it is stated that (1992):

> In order to protect the environment, the precautionary approach should be widely applied by States according to their capabilities. Where there are threats of serious or irreversible damage, lack of full scientific certainty shall not be used as a reason for postponing cost-effective measures to prevent environmental degradation.

Argumentative versions of the precautionary principle are often grounded in narrow utilitarian ethics and include economic evaluation, i.e. risk/cost-benefit analyses of environmental risk or evaluation of the cost-effective nature of protection of the environment. In this context, the precautionary principle may be used as an option to manage risks after they have been identified through risk analysis. When applying risk/cost-benefit analyses, the risks and the benefits need to be identified, and the probabilities for risks are assessed by employment of statistical evidence or quantified by observable frequencies before they are taken into account (Karlsson 2006). However, full certainty about potential unintended effects is unattainable prior to use of new technology. The complexity and the variability of ecosystems make it impossible to give estimates that can be used in risk/cost-benefit analyses. Hence, application of argumentative versions of the principle may mean that potential risks to health and the environment are downplayed when there is a lack of scientific understanding and/or scientific disagreement about the relevance of the potential adverse effect (Myhr and Traavik 2007).

Prescriptive versions of the principle embrace inherent values of the environment and often are founded on ecocentric views or duty-based concerns for nonhuman beings and ecosystems. The Wingspread Statement (Raffensperger and Tickner 1999) is considered to represent an prescriptive version of the precautionary principle; it states:

> When an activity raises threats of harm to human health or the environment, precautionary measures should be taken even if some cause and effect relationships are not fully established scientifically.

Another prescriptive version is the Bergen Declaration (1990), which includes precaution as an essential tool for sustainability:

> In order to achieve sustainable development, policies must be based on the Precautionary Principle, environmental measures must anticipate, prevent, and attack the causes of environmental degradation. Where there are threats of serious or irreversible damage, lack of scientific certainty should not be used as a reason for postponing measures to prevent environmental degradation.

Prescriptive versions of the precautionary principle are active in nature and commit regulators to take action by (a) implementing risk management procedures such as, for instance, monitoring and/or (b) shifting the burden of proof to the proponent or developer to increase the scientific understanding. GMOs offer beneficial prospects for health and the environment on a short-term scale, but there are indications of long-term ecological costs. Hence the prescriptive versions of the principle seem to be the most suitable to minimize ecological risk. This ultimately leads to the issue of how to properly deal with the scientific uncertainties surrounding GMOs.

6.3 The Precautionary Principle and GMO Research

When discussing risks and uncertainty in connection with GMO use and release, the first step is for scientists to become conscious of the role they play in producing information and the subsequent political use of this information. The responsibility of scientists has been well reflected in the traditional methods of pursuing and communicating discoveries in science and by exposing the results to objective scrutiny by the scientific community. More recently, it has been argued that scientists also have an obligation towards the environment and society, which is not limited to conducting research, but implies that scientists also have to consider social and ethical implications of research products when initiating research (Melo-Martin and Meghani 2008, Mayer and Stirling 2004). While conducting research, scientists make assumptions, choices, and inferences based on professional judgment and standard practices, which, if unknown to the public or policy-makers, may make scientific results appear more certain and less value-laden than warranted. For example, during observational studies of complex, poorly understood systems, errors in the independent variables, errors arising from the choice of the wrong form of model being used to analyze and interpret the data, and bias from the conduct of the study may arise (Kriebel et al. 2001). Experiments with a handful of replicates and low statistical strength could fail to expose risks associated with a particular activity. Under the assumption "safe until proven otherwise" the regulator may fail to boost the rigor of testing, and weak indications of harm may be considered insignificant.

6.3.1 Hypothesis Testing: Type I-Errors Versus Type-II Errors

According to the traditional scientific norm, one ought to have complete and supportive information before claiming a cause–effect relationship. Consequently, the statistical significance of the result must be strong enough to allow only a small probability (p) that the result is due to chance or has been based on speculation. By convention, in a Type-I setting the probability of this error being made is determined by the significance level of α – often at 5%. Hence, if there is less than 95% confidence that there is an effect (1 in 20), H_0 (the null hypothesis) is not rejected (Lemons et al. 1997). In such situations, scientists are prone to assume that the evidence is not strong enough to reject H_0, that there is no effect due to limitations of the experimental design or test, or that the data variables are too similar to detect an effect. Minimizing Type I-errors is necessary and adequate when doing laboratory research, as the parameters and variables are few, the results are in most cases reliably identifiable or quantifiable, and the purpose is to obtain new understanding and avoid spurious results.

However, in risk-related research, standards and burden of proof have to be differentiated from those used in fields where there is high certainty and high consensus among scientists. Complex interactions in open systems cannot be adequately predicted; hence achieving complete and supportive information before claiming a cause–effect relationship may not be possible. This means that risks to society, to health, and the environment may remain obscured, because a bias towards avoiding Type-I errors discourages research into risk-associated aspects.

6.3.2 Type-II Errors and Early Warnings

The report "Late Lessons from Early Warnings: the Precautionary Principle 1896–2000," published by the European Environment Agency (Harremoës et al. 2001), clearly illustrates how narrow risk assessment frameworks and the choice of null-hypothesis affect initiation of early warning research. In the report 14 cases are described, and the conclusion is that not taking precautions may have human, ecological, and economic costs, and the authors stress the need to learn from past failures and to heed early scientific evidence of risks. This means that risk to the society and the environment may be increased because a bias towards avoiding Type-I errors discourages risk-associated research and communication of early warnings. Accordingly, given the asymmetry in the consequences according to the choice of avoiding Type-I versus Type-II errors, application of a precautionary approach would entail a research agenda seeking to identify potential threats to health and the environment before they become real.

Controversial GMO studies have illustrated that early warnings need to occur within an existent scientific frame of reference to avoid them being considered insignificant and/or negligible (Myhr 2005). Further, the ability of early warnings to effect change depends on the vested interests involved and the availability of systems that identify and respond to unexpected findings. Many of the proponents of GMO use and release have economic or political interest in the development of GMOs, hence causing an asymmetric relationship of power between proponents, skeptics, and opponents (Sagar et al. 2000). The most worrying aspect at present is that only a fraction of GMO research is directed toward the investigation of potential harm to health and the environment by GMO use and release.

6.3.3 Contextualization of the Broader Scientific Uncertainties

There is a growing awareness that quantitative methods might leave out many important aspects of uncertainty (see for instance Funtowicz and Ravetz 1990, Stirling

2007, Walker et al. 2003, Wynne 1992). For instance, uncertainty with regard to GMO release and use can be presented at different levels:

- Uncertainties that refer to situations where we do not know or cannot estimate the probability of hazard, while those hazards are known. This may be due to the novelty of the activity, or to the variability involved.
- Ignorance that represents situations where the kind of hazard to be measured is unknown, i.e. completely unexpected hazards may emerge. This has historically been experienced for instance with BSE (Bovine Spongiform Encephalopathy), dioxins, and pesticides (Harremoës et al. 2001). With regard to GMOs unprecedented and unintended nontarget effects may emerge. Nontarget effects include the influence on and interactions with all organisms in the environment, and may be direct or indirect. Direct effects concern ecotoxic effects on other organisms. Indirect effects concern effects on consumer health, contamination of wild gene pools, or alterations in ecological relationships.

Employment of model-based decision support, such as for instance the Walker & Harremöes (W&H) framework (Walker et al. 2003), may help to identify the types and levels of the uncertainty involved. The W&H framework has been developed with the purpose of providing a state-of-the-art conceptual basis for the systematic evaluation of uncertainty in environmental decision-making. One of the main goals of the W&H framework is to stimulate better communication between the various actors in the identification of areas for further research and in decision-making processes. In this framework, uncertainty is recognized in three dimensions:

- Location [where the uncertainty manifests itself, (e.g. distinction between context (ecological, technological, economic, social, and political), expert judgment and considerations, models (model structure, model implementation, data, outputs etc.)],
- Nature (degree of variability which can express whether uncertainty primarily stems from inherent system variability/complexity or from lack of knowledge and information),
- Level (the severity of uncertainty that can be classified on a gradual scale from "knowing for certain" to "complete ignorance").

Krayer von Krauss et al. (2004) have tested the W&H framework with the purpose of identifying scientists and other stakeholders' judgment of uncertainty in risk assessment of GM crops. In these studies the focus was on potential adverse effects on agriculture and cultivation processes by release of herbicide-resistant oilseed crops. Krayer von Krauss et al. interviewed seven experts in Canada and Denmark. To identify the experts' views on location uncertainty the authors presented a diagram showing causal relationships and key parameters to the experts. With the purpose of identifying the level and nature of uncertainty, the experts had to quantify the level and describe the nature of uncertainty on the key parameters in the diagram. By asking the experts to identify the nature of uncertainty it was possible to distinguish between uncertainty that may be reduced by doing more research and ignorance that stems from systems variability or complexity.

Approaches that define and systematize the uncertainty involved, such as the W&H framework, may help to use scientific knowledge more efficiently, in directing further research and in guiding risk assessment and management processes.

6.4 Complex Interactions by GMO Use and Release

GMOs are released into and interact with a complex biological and social system. Complex systems are generally described as systems that can be explained in several plausible ways. They are open, and dependent on several dynamic and nonlinear output–input interactions. Thus, changes in initial conditions can have pervasive and unpredictable effects, or novelties may appear. The difficulties in making predictions become especially evident when taking a long time frame and wide geographical range into consideration.

For transgenic GMOs the complex interactions may be illustrated at several levels:

- Genomic level: The DNA structure of the organism that is being modified is a complex structure, and there is a lack of understanding as to how and where a transferred gene integrates with the organism's genome, and if there are interactions between the introduced transgenes and the other genes in the genome (Doerfler et al. 2000, Latham et al. 2006).
- Organismal level: A GMO released into the environment will interact with the other organisms in this environment. For example, the intention of insect resistance in Bt maize is to have an impact on the pest population of insects feeding on the plant, but it may also impact nontarget organisms, as other insects feeding on the plant, some of which may even be beneficial, as well as other consumer organisms of the maize (Bøhn et al. 2008, Rosi-Marshall et al. 2007).
- Population level: Release of a GMO can cause changes in the natural population of other organisms. The modified organisms can cross with native species and change the gene pool of that population, possibly causing a change in fitness of the hybrid population.
- Ecosystem level: This can include tri-trophic impacts, like impacts on insects, birds, and animals feeding on insects that are feeding on GMOs (Hilbeck and Schmidt 2006, Lövei and Arpaia 2005). It also includes the impact on consumer health, impacts on soil construction and the possibility of horizontal gene transfer as genes are released through degradation of decaying GMOs.
- Ethical and social implications: Implications for the development of agriculture and maintenance of local agricultural practices, small-scale agriculture and local plant varieties (which often are aspects of local culture etc.), and socio-economic impacts on local agriculture societies. Implications for the public's perception of food and ethical considerations as to how humans interfere with other species and the environment.

When GMOs are released into complex environmental systems, delayed harm might occur, it may cause irreversible changes, and since GMOs are reproductive living organisms, changes might increase instead of decrease over time. Hence, some results based upon reductionistic assumptions, such as the belief that possible large-scale effects from GMOs can be extrapolated from effects studied in small-scale models, do not have validity and do not represent reality. To extrapolate from one context to another, i.e. from small to large-scale release, leaves questions unanswered concerning the scale of effects and therefore the environmental fate of a GMO (Wolfenbarger and Phifer 2000, Haslberger 2006). Environmental conditions are geographically and climatically different throughout the world and this may therefore make it difficult to identify the cause–effect relationships of impact, especially when using studies conducted in a different environment as the basis for risk assessments. Such extrapolations may in fact increase the uncertainty. This is also because uncertainties regarding the behavior of complex systems may not be directly linked to lack of knowledge which can be reduced by performing more research.

Designing adequate human and environmental models for determination of risks and identification of unpredictable effects are difficult tasks. Present approaches need to be supplemented with methods that prescribe broader assessments and study whole systems. This involves a perception that the dynamics of human and environmental systems cannot solely be described by its parts, such as genes and proteins, but concern interaction with each part of the system. Developments within functional genomics and molecular biology such as genomics, proteomics, and metabolomics are advancing and represent new toolboxes that can be used to gain new scientific knowledge. Furthermore, as GMOs are to be used and released into the environment there is a need to supplement molecular biology approaches with other approaches that facilitate explanation and understanding of complex ecological issues.

This makes it necessary to involve a wide base of scientific disciplines as well as independent scientific institutions in the pursuit of scientific understanding. Hence, the different methods and models representing the different disciplines may be seen as compatible providers of information and models for studying the problem or the system. With more diversity in the approach, more data will be generated and more responses will be available to understand complexity and changing conditions.

6.5 Scientific Dissent with Regard to Impacts from GMO Use and Release

In risk assessment and risk management processes lack of scientific understanding may create expressions of different viewpoints with regard to the scope of risk frameworks and to the significance of potential ecologically adverse effects. There are for instance divergent opinions about the definition of potentially "adverse effects," about the relevance of various potentially adverse effects, and what action to take to prevent adverse effects (Myhr and Traavik 2003). Sarewitz (2004) argues that scientific dissent in the case of highly complex and difficult-to-assess risk situations are due to different backgrounds/disciplines. The scientists' backgrounds may affect choice of hypotheses, methods and models, which give conflicting data and cause disagreements among scientists. Scientists make assumptions based on the paradigm they have trained under and incentives provided within their current work situation. The importance of values and interests influencing the scope of the research, both in problem framing, choice of methods, and in the interpretation of data and communication of findings must be recognized. It has also been shown that different scientists, partly depending on whether they work within industry, government or academia, interpret data differently in situations characterized with uncertainty, and thus express a diversity of opinions on the issue (Kvakkestad et al. 2007). To some extent, competing claims to knowledge are also produced by the NGO (Non-Governmental Organization) sector, and publicized widely in the media. Hence, there are competing knowledge claims, stemming from different groups, organizations, research institutions, and stakeholders. From this perspective, the demand for "more research" is not sufficient to reduce scientific uncertainty, since the incapacity of science to provide a unified picture of the environment contributes to the uncertainty.

Yet, keeping in mind the subjective context of scientific practice and data production, few would not agree that continued research on biosafety issues would contribute to improving the safe use of GMOs. The lack of scientific consensus is a normal and often the driving part of science and is not a particular risk feature of biotechnology GMOs. Sarewitz (2004) denotes this observation as an "excess of objectivity" referring to the observation that available scientific knowledge can legitimately be interpreted in different ways to yield competing views of the problem and therefore differences in society's response. Meyer et al. (2005) argue that the current lack of data and the subjective constituents, particularly integral values, within data production in biosafety hinder scientific consensus building on the effects caused and observed. Moreover, a nonuniform response is seen among experts to new studies reporting deviations from safety assumptions further exemplifying the values, stakes, and subjective interpretations underlying the discourse on the safety of GMOs.

Following Sarewitz (2004), there is a need for enhanced communicationS on the plurality of perspectives and implicit values in science. Reflecting on the role of scientific disunity in the interpretation of scientific uncertainty related to GMO use and release, the question arises as to how enhanced dialogue between competing disciplines can contribute to making explicit those values, interests, and implicit assumptions that represent the frame for each discipline's approach to scientific uncertainty. For instance, involvement with a wide basis of scientific disciplines will ensure diverse consideration of both mainstream and minority opinions, and cause avoidance of abuse of science by scientists biased toward a specific agenda. This will not reduce controversy but may, if the different scientists are open to mutual learning, create new ways to conduct and organize research, to identify ways to bridge the different conceptual differences and to engage in meaningful cross-disciplinary communication.

6.6 Normative Concerns with GMOs

The case of GMOs in agriculture has been illustrative of the types of problems that may emerge when the appropriate scientific and normative standards cannot be agreed upon. Risk assessment and management strategies are developed within particular frameworks, including normative standards and preferences regarding our relation to the natural environment and the preservation/promotion of human health. For instance, in the EU directive 2001/18/EC it is stated that an environmental risk assessment needs to consider direct and indirect effects, immediate and delayed effects, as well as potential cumulative and long-term effects due to interaction with other GMOs and the environment. Article 1 of the Cartagena protocol specifies that the entire objective of the document is to protect and conserve biodiversity according to a precautionary approach. One of the purposes of the Norwegian Gene Technology Act is that use of GMOs shall be in accordance with the principle of sustainable development. Hence, the data generated from a risk assessment must fit into a normative framework, and need to reflect the expressed criteria for approval. Normative standards may affect the scope of risk management of GMO use and release, and affect legal interpretations concerning the acceptable risks, and thereby function as guidance for when and how to apply the precautionary principle. However, interpretation of the normative standards that shall be used to guide risk assessment and management of GMOs is still an unresolved issue. Undoubtedly, participative processes are needed to identify perspectives on the ethical, social and legal interpretations of the normative standards that are found in the different GMO regulations.

The most prominent mistakes related to emerging technologies have been the presumption that the reason for public skepticism was lack of information and/or scientific illiteracy (the deficit model). Another mistake was the overreliance on technical aspects of risk assessments and credible scientific backing by experts. A third mistake was the strong belief that technological fixes were the most appropriate and "efficient" solutions to societal problems. As a result, this technocratic approach has largely hindered real integration of the ethical, social, and legal aspects of new technologies, and its concerned constituencies. With the intention to avoid these mistakes from the past, the ultimate challenge becomes how to promote new approaches that allow the generation of suitable working paradigms that adequately address broader concerns (Haslberger 2006).

One solution may be approaches that allow for a coordinated forum for considerations of societal, ethical, and legal questions concerning the introduction of new technologies, which involves debates on technological developments and that reduce the time lag, distance, and power asymmetry between "upstream" and "downstream" innovation processes (Wilsdon et al. 2005, Wynne and Felt 2007).

This will require better integration into biotechnology research and development, a kind of conscientious dialogue and framework building with multi-stakeholder involvement. Such approaches can help to ensure that future GMOs are more sustainable and socially relevant, and ultimately improve public confidence in that normative standards are implemented in assessment and management processes of new biotechnologies.

6.7 Conclusion

The precautionary principle provides a basis for adequate consideration of ecological and ethical issues of vital importance to the protection of health and the environment from unforeseen adverse effects of GMOs. This challenge has to be met by scientific conduct and approaches that aim to:

- Manage risk and uncertainty, taking into account uncertainty, ignorance, and complexity of the ecological systems that the GMOs are to be released into.
- Improve the framework and context of the methodological design of health and environmental studies to ensure scientific and social robustness of the knowledge generated. In this context, scientists and decision-makers should become comfortable with making decisions based on the weight of evidence according to an approach that strives to reduce Type-II errors.
- Realize that different normative conceptualizations of frame affect the scope of research and the significance of the final outcome.

These points are of great importance for the implementation of precaution in risk management and for the initiation of precautionary-motivated research which is needed so that scientific understanding can be used as the basis for such management.

Bibliography

Anne Ingeborg Myhr is employed as a senior scientist at Genøk – Centre of Biosafety in Tromsø, Norway. She has a master in biotechnology from NTNU, Trondheim and a PhD from the University of Tromsø. Myhrs present research engagements are within risk governance of the use of GMOs, DNA vaccines and nanobiotechnology. Myhr is also involved in capacity building in biosafety of GMO use and release in the third world.

References

Bøhn T., Primicerio R, Hesssen DO, Traavik T (2008) Reduced fitness of Daphnia magna fed a Bt-transgenic maize variety. Arch Environ Contam Toxicol 55:584–592
CBD. (1992) Cartagena Protocol on Biosafety (http://www.biodiv.org/biosafe/protocol)

European Council Directive 2001/18/EC (http://www.europa.eu.int/commm/food/fs/sc/scp/out31_en.html)
Doerfler W, et al. (2001) Foreign DNA integration – Perturbations of the genome–oncogenesis. Ann NY Acad Sci 945:276–288
Funtowicz SO, Ravetz J R (1990) Uncertainty and quality in science for policy. Kluwer, Dordrecht, pp. 7–16
Gene Technology Act (1993) The Act relating to the production and use of genetically modified organism. Act no. 38 of 2 April 1993, Oslo, Norway
Harremoës P, Gee D, MacGarvin M, Stirling A, Keys J, Wynne B, Vaz SG (2001) The precautionary principle in the 20th century. Late lessons from early warnings. Earthscan Publications Ltd., London (http://reports.eea.eu.int/environmental_issue_report_2001_22/)
Haslberger AG (2006) Need for an "integrated safety assessment" of GMOs, linking food safety and environmental considerations. J Agric Food Chem 54:3173–3180
Hilbeck A, Schmidt JEU (2006) Another view on Bt proteins – How specific are they and what else might they do? Biopestic Int 2:1–50
Karlsson M (2006) The precautionary principle, Swedish chemicals policy and sustainable development. J Risk Res 9:337–360
Krayer von Krauss M, Casman E, Small M (2004) Elicitation of expert judgments of uncertainty in the risk assessment of herbicide tolerant oilseed crops. J Risk Anal 24:1515–1527
Kriebel D, et al. (2001) The precautionary principle in environmental science. Environ Health Perspect 109:871–876
Kvakkestad V, Gillund F, Kjølberg KA (2007) Scientists' perspectives on the deliberate release of GM crops. Environ Val 16:79–104
Latham JR, Wilson AK, Steinbrecher RA (2006) The mutational consequences of plant transformation. J Biomed Biotechnol 6:1–7
Lemons J, Shrader-Frechette KS, Cranor C (1997) The precautionary principle: Scientific uncertainty and type I and type II errors. Found Sci 2:207–236
Lövei G L, Arpaia S (2005) The impact of transgenic plants on natural enemies: A critical review of laboratory studies. Entomol Exp Appl 114:1–14
Mayer S, Stirling A (2004) GM crops: Good or bad? EMBO Rep 5:1021–1024
Melo-Martin I, Meghani Z (2008) Beyond risk. EMBO Rep 9:302–308
Meyer G, Folker AP, Jørgensen RB, Krayer von Krauss M, Sandø P, Tveit G (2005) The factualization of uncertainty: Risk, politics, and genetically modified crops – A case of rape. Agric Hum Val 22:235–242
Myhr AI (2005) Stretched peer-review of unexpected results (GMOs). Water Sci Technol 52:99–106
Myhr AI, Traavik T (2003) Genetically modified crops: Precautionary science and conflicts of interests. J Agric Environ Ethics 16:227–247
Myhr AI, Traavik T (2007) Poxvirus-vectored vaccines call for application of the precautionary principle. J Risk Res 10:503–525
Raffensperger C, Tickner J (eds) (1999) Protecting public health and the environment: Implementing the precautionary principle. Island Press, Washington, DC
Rosi-Marshall EJ, Tank JL, Royer TV, Whiles MR, Evans-White M, Chambers C, Griffiths NA, Pokelsek J, Stephen ML (2007) Toxins in transgenic crops byproducts may affect headwater stream ecosystems. PNAS 104:16204–16208
Sagar A, Daemmrich A, Ashiya M (2000) The tragedy of the commoners: Biotechnology and its public. Nat Biotechnol 18:2–4
Sandin P (2004) The precautionary principle and the concept of precaution. Environ Val 13:461–475
Sarewitz D (2004) How science makes environmental controversies worse. Environ Sci Policy 7:385–403
Stirling A (2007) Risk, precaution and science: Towards a more constructive policy debate. EMBO Rep 8:309–315

Walker WE, et al. (2003) Defining uncertainty: A conceptual basis for uncertainty management in model based decision support. J Integr Assess 4:5–17

Wilsdon J et al. (2005) The public value of science; or Howe to ensure that science really matters. http://www.demos.co.uk/catalogue/publicvalueofscience/

Wolfenbarger LL, Phifer PR (2000) The ecological risks and benefits of engineered plants. Science 290:2088–2093

Wynne B (1992) Uncertainty and environmental learning: Reconceiving science and policy in the preventive paradigm. Global Environ Change 2:111–127

Wynne B, Felt U (2007) Taking European Knowledge Society Seriously. European Communities: Directorate-General for Research Science, Economy and Society

Chapter 7
Biotechnology, Battery Farming and Animal Dignity

Peter Kunzmann

Abstract In combination with the question of the legitimacy of biotechnology applications on animals, the author discusses the extent of the concept of "animal dignity" and confronts it with other basic controversial concepts of animal ethics. After discussing the core idea and applicability of this concept of animal dignity, he shows that adequate treatment of animals is contradictory to biotechnological alteration of animals. Then he gives reasons why the concept of dignity should lead the debate on animal ethics: The concept posits the guarantee of animal welfare as well as human responsibility for animal treatment.

7.1 The Success of a Concept

A few months ago the Swiss Federal Ethics Committee on Non-Human Biotechnology (known to experts as the EKAH) and the citizens of Switzerland were awarded the spoof "IgNobel Peace Prize 2008" for "adopting the legal principle that plants have dignity".

Is it so funny to speak of dignity when referring to living beings other than humans?

This may be the case with plants – but how about animals? Is it so ridiculous to ascribe some kind of dignity to them? As early as 1978 UNESCO used the term dignity in the Declaration of Animal Rights.[1] But when translating this text into Polish, the Polish Ethological Society added a genuine warning, saying that here UNESCO authors went definitely astray: The use of the term dignity is to be restricted to

---

P. Kunzmann (✉)
Ethikzentrum Jena, Jena, Germany
e-mail: peter.kunzmann@uni-jena.de

[1] Article 10b: Animal exhibitions and shows that use animals are incompatible with any animal's dignity.

humans only.[2] One may be sure that these Polish scientists are not alone in their reluctance to broaden the use of a powerful word beyond "human dignity".

Whether UNESCO actually was ill-advised to speak of dignity in this given context can be left aside for a moment. The phrase "dignity of animals" must nevertheless be considered a highly successful coinage. Twelve years ago, Bressler (1997) stated that the German word for dignity, Würde, is hardly to be found in connection with animals, obviously due to its strong relation to human dignity.[3] In the meantime, this has changed considerably: dignity/Würde has become widespread in use, whether it is part of public rhetoric or as the subject of serious ethical and legal debates.

This success must have its reasons. To mention a few[4]:

Dignity and its equivalent Würde are somewhat catchy formulas that provide a certain contrast to more technical terms like "integrity" or to phrases like Mitgeschöpflichkeit, which is meant to be the basic principle in the German Animal Protection Law. This term is so uncommon, so artificial and so uneasy to handle that it has never made it into normal conversational German. It is evidently less attractive than dignity.

At the same time, dignity conveys an almost religious importance, which may make it attractive to those who use the term, and which is at the same time one of the reasons why other ethicists refrain from using it. This applies not only to animal dignity, but to all applications of the term as an ethical principle. There is a fierce debate in the US about whether human dignity is a helpful concept in Medical Ethics. In 2008 the "President's Council on Bioethics" for example edited a 555-page-report on the topic "Human Dignity and Bioethics". This large volume is just one example that reflects the controversy on the use of the term. It also contains statements on the necessity of speaking of dignity as well as fierce opposition to the statement. Several authors refer to an article by Ruth Macklin who argues that dignity is a useless concept. Everything that it attempts to convey is already expressed by "autonomy" or by "respect for the person's autonomy".

The quarrel nonetheless shows clearly that even the opponents know that dignity is a powerful term that indicates serious interests.

Dignity has also a very positive connotation which makes it even more attractive to speak of when propagating animals' concerns in public – it has, as J-C Wolf once put it, a "manifesto" character. A manifesto character may mean that speaking of dignity states something important which at the same time urges some kind of specific action or attitude. In terms of speech act theory, locutionary acts containing dignity tend to perform some kind of warning, of declaring, of stating, of

---

[2]"Uwaga: tu prawodawcy się zagalopowali – godność jest atrybutem człowieka" (http://www.nencki.gov.pl/ptetol/prawa_zwierz.pdf).

[3] "Es gibt (...) noch sehr wenige philosophische Beiträge zur 'Würde des Tieres'. Dies mag seinen Grund darin haben, daß, wenn von Würde die Rede ist, die 'Menschenwürde' gemeint ist und dieser Begriff eben zur Abgrenzung und Auszeichnung des Menschen gegenüber dem Tier dient". (Bressler 1997, p. 191).

[4] See also Kunzmann (2007, pp. 16–18).

demanding, of protesting, rather than just describing something. Dignity is kind of a so-called "thick concept" which at the same time carries a description of a state of affairs AND an evaluation of the depicted state of affairs, usually a critical one. This, at least, applies to the long history of human dignity, which in many cases was used as a protesting manifesto against circumstances which were considered to violate this moral and legal principle. Although it is rather difficult to give human dignity its precise shape, it is easier and therefore more common to protest whenever human dignity is at stake.

Animal dignity may borrow some of the vigour from human dignity, but it pays a price. It states something; it demands something without explaining it too clearly.

And to some of the critics of the concept of "dignity of animals" or "dignity of creatures", there is another danger lurking. The refutation of transferring dignity from humans to animals would broaden the term beyond its original meaning without cause; this refusal comes most often from authors specializing in law or legal philosophy (Löwer 2001, von der Pfordten 2003). There is a fear that the core concept of dignity will suffer when we apply it to entities which we treat in a way that is incompatible with dignity: Humans kill animals, some for quite minor reasons. Animals are kept in confinement, are bought and sold, and are the subject of all sorts of treatment that would absolutely not be justifiable for humans. How can we at the same time ascribe dignity to animals and do all those nasty things which we would refuse to do to human beings with respect to human dignity? As long as our common social practice allows humans to act with such licence with respect to animals, it is not possible to attribute a dignity to them that would exclude these practices when performed on human beings. Will, in turn, the use of the same concept lead to a softening and blurring of its content for humans?

7.2 The Need for a New Concept

The Swiss first invented the phrase "dignity of creatures"[5] and then they subsequently deleted the phrase from their constitution – at least in the French version of 2000. They replaced the phrase "dignity of creatures" with "integrity of living beings". Though it came as no surprise, not all the experts involved were happy about this development.[6] According to Richter (2005), "dignité de la creature" sounds pointless and even ridiculous in francophone ears.[7]

It is not at all clear how the two notions of dignity are related. Is dignity of animals the same as human dignity? Or is it just a minor version of it? What then should it mean? All these semantic problems (which were debated in ethicists' discussions)

---

[5] Baranzke (2002) has made a full report on the process that resulted in the article and also provides in-depth analysis of its legal and ethical impact.

[6] See EKAH (2000): Stellungnahme zur französischen Version des Art. 120 BV. Bern.

[7] "Die infrage kommende Formulierung 'dignité de la créature' klang... in französischen Ohren... 'unglücklich' und 'lächerlich' [sic!]" (Richter 2005, p. 348).

did not hinder the success of the formula at hand. But even its certain vagueness seems not really problematic. On the contrary, seen from a pragmatic point of view, all this has made it possible to use the phrase in a huge variety of contexts and in connection with many different issues. The dignity of animals has, in fact, proven to be quite versatile.

Let us recall the occasions where this so-called dignity of animals is referenced. UNESCO, as mentioned above, applied it to the situation of animals being fairground attractions for human amusement. In itself, this does not necessarily harm animals nor does it inflict pain upon them. But there may still be moral concerns regarding the use of animals as show stars. The same applies to animals put into strange outfits or in positions meant to be funny, for example when animals appear in human dress[8] which may be the case in advertising commercials. It does not hurt them apparently, but it still causes a feeling of unease in some people.

In Switzerland, it was the political debate over genetic modifications that led to the remarkable passage in the Swiss constitution (which in turn led to the remarkable EKAH statement mentioned above and the remarkable international award). In this important article (now article 120), the Swiss constitution obligates legislation to respect the "dignity of creatures" when regulating biotechnology. Although biotechnology does not necessarily inflict pain upon animals, it seems highly appropriate to limit human efforts to modify animals (and plants) merely on the basis of their desires. The Swiss framed it with the term "Würde". Legislation in some countries bans the procreation of animals, whether by traditional breeding or by the use of gene technology, whenever the resulting animals would suffer under their inherited (traits?) properties. Should we in turn suppose that anything else is simply morally correct? Are completely naked (furless) kittens acceptable as long as it does not affect their overall welfare? How about gigantic but healthy salmon? How about cows that are bred never to have horns (Kunzmann 2005)? And how about chickens without feathers as they were introduced in 2001?

The intuitions of many Westerners would suggest: Rather not. This seems to be a widespread moral opinion within Western societies. But this opinion seems hard to explain and maintain on purely ethical grounds. Manuel Schneider (in his online article on the dignity of creatures) has given a similar list, and he added a few more applications of the principle (see also Schneider 2001). One may read such lists[9] as a summary of unsolved problems in animal ethics, but one common feature seems to be that these cases cannot be settled with the "pathocentric" principles of modern animal welfare alone. It seems insufficient to intervene only to avoid or diminish animal suffering.

---

[8] See for example EKAH/EKTV (2001, p. 6).

[9] See another one for example in Steiger (2002, p. 229).

7.3 Core Ideas

This may be enough to explain the enormous success of the phrase which in my opinion is due to its usefulness. A concept proves successful when it expresses something new and at the same time desirable. One might say: To prosper and grow within a language, a new word needs a kind of ecological niche, and "dignity of animals" has found it. Using the form of a Wittgensteinian (PI § 257) metaphor one could say that there is already a place in language prepared for the word that comes into practice. In the case of dignity with regard to animals I am inclined to say that this place can be pinpointed accurately, and Brom (2000, p. 61) has found its perfectly matching name when talking of "protection beyond pain and injury".[10]

I consider this to be the common feature in all those uses of the word: It is all about animal protection beyond pain and injury. Let me give you an example:

> [I]f somebody suggested genetically engineering a race of mentally deficient human beings who would not mind being enslaved and mistreated by us, most of us would certainly find the very idea abhorrent and morally detestable. Now why should the same reaction not be appropriate with regard to animals? Take, for instance, the situation when a bird is kept in a small cage where it cannot fly. Maybe it suffers, we imagine it probably does. But the point is that even if it did not suffer, and we knew it did not, we would still feel the wrongness of keeping it in a cage. And again, it would somehow make the whole thing even worse, if we had found a way to manipulate the bird so that it did not even want to fly anymore.

I quote this passage from Hauskeller at large because it draws an essential parallel between the violations of human dignity and violations of the dignity of animals. In the case of humans being bred as slaves – even without suffering – the term dignity comes instantly to mind. And although Hauskeller does not refer to it when describing the situation of the animals, he gives a perfect example of violations of what could be called the dignity of animals. His example also shows how closely the topics in this chapter, as they appear in the title, are connected. The dignity of animals could be considered a direct answer to new challenges in animal ethics – animal biotechnology on the one hand and battery farming on the other. In both cases animal suffering does not appear to be a sufficient reason for ethical concerns. Laying hens kept in cages probably suffer from being deprived of the opportunity to perform their natural ethological programme; but even though they show no sign of distress, a large part of the human population in western countries is strictly opposed to this form of animal husbandry. What are we going to make of breeding animals to fit better into these "unnatural" circumstances? Breeding is one way, biotechnology is another one. In both cases the animals are altered to meet the conditions under which they are supposed to live. It may be, of course, that a certain type of breeding or the modern use of genetic information leads to an improved way of life for the animals. One case in which both perspectives meet is stress reduction in pigs. Some 10 years ago, pigs were specially bred for high stress tolerance. Here the adaptation

---

[10] Although I must admit that Brom had something much more specific in mind, his words hit the mark.

of animals seems to be aimed at reducing suffering under given conditions. But is it "fair" to change the "nature" of pigs to fit into an environment that is perfectly shaped according to the needs of their human owners? Animals may get away unhurt but certain human actions may be condemned nonetheless. In these cases, the plea for dignity is not far away. [11]

There are plenty of occasions where our term could apply. Those mentioned in the title of this article, namely biotechnology and battery farming, are the most prominent. They may come together, like in the use of modern genetics for breeding farm animals. Only science fiction has depicted this in its most grotesque forms: a pig specially designed to be happy when it is eaten and still capable of saying so, is the imaginary invention of the eccentric author Douglas Adams for his "Restaurant at the End of the Universe". Scientifically more palatable but still science fiction is the idea of the so-called AML (Schmidt 2008, p. 43). AML, or "animal microencephalic lumps" are animals that have been modified to the extent that they finally feel neither pain nor desire; they have been engineered to exclude any biological functioning unnecessary for the production of desirable goods such as eggs, meat and such.

For us to be concerned it is not a question of the means. In the case of animal husbandry it does not really matter whether a fundamental change in the general constitution of animals was achieved by "good old breeding" or by the use of biotechnology; though the latter resulted in more controversy and thus led to the article in the Swiss constitution. The conflict is also not confined to this type of human–animal relationship. The problem has the same structure, for example, in the case of so-called knockout-mice, genetically manipulated for experiments. They may be better off than other animals used in laboratories. A comparatively long tradition in animal ethics for these situations is the so-called 3R principle, of replacing, reducing and refining. According to this principle, animals are genetically modified for their use in the laboratory, thus refining the experiments and reducing animal suffering. From this perspective, modifications may not only be legitimate, they may appear preferable or even obligatory. Quite like the farm animals that are adjusted to husbandry conditions according to the overriding interests of humans, laboratory animals could be adjusted to the circumstances, much to their own advantage.

As I mentioned there are serious ethical questions that cannot be answered by just referring to animal welfare.

---

[11]"'Stressresistenz' und damit implizit Leidensvermeidung ist z.B. in der molekularen Schweinezucht eines der zentralen Zuchtziele. Und dennoch scheinen prima facie gerade Eingriffe am Erbgut und an der genetischen Integrität der Tiere unter ethischen Gesichtspunkten problematisch zu sein. Diese offensichtliche Lücke innerhalb der Tierschutzethik und des Rechts galt es zu füllen, und die Einführung des Schutzzieles 'Würde der Kreatur' hatte zunächst diese Funktion". (Schneider 2001, p. 229).

7.4 Integrity

One way of answering these questions was the introduction of the principal of "integrity". Protecting and maintaining integrity means to preserve an organic completeness – with no regard to how diminishing this completeness may be counted subjectively in this very organism. In other words, to reduce or harm the integrity of an organism is ethically relevant, even if animals or plants do not feel anything from it (are not adversely affected?).

This theory has been prominently advocated in the Netherlands (Ferrari 2008, p. 168) where criteria for it were determined. Integrity comprises three different dimensions: "A valuable interpretation of respect for the principle discussed here is that recently advanced by Rutgers and Heeger (1999)... Thus, they suggest that respect for animal integrity comprises three elements, all of which need to be satisfied for integrity to be respected". The elements are "(1) the wholeness and completeness, (2) the balance in species specificity, and (3) the capacity (for an animal) to independently maintain itself" (Mepham 2000, p. 70).

In Heeger's words (2000, p. 50): "One could consider adopting a moral principle of respect for animal integrity, saying that we ought to refrain from attacking the intact body of animals, from threatening the achievement of the ends and purposes characteristic of them, and from depriving them of the capacity to maintain themselves in an environment appropriate for them".

At first sight, the concept of integrity solves quite a number of problems:

First of all, it avoids any confusion with the somewhat "absolute" concept of human dignity. This may be one of the reasons, why in the year 2000 the Swiss translation service changed the text of the Swiss Constitution quite considerably. The whole constitution was reconstructed, and when it came to the French translation of the text (Richter 2005) "dignité de la créature" was replaced by the phrase "intégrité des organismes vivants" (Ferrari 2008, p. 169). The EKAH reacted immediately (EKAH 2000); one of the arguments was that the term integrity poses as many problems as the term dignity does: "It is not clear whether the concept of integrity refers to integrity in its physical-biological, its psychic or its physical sense". (Es ist ungeklärt, ob sich der Begriff [intégrité] auf eine physisch-biologische, auf eine genetische oder auf eine seelische oder metaphysische Integrität bezieht.) In the Dutch discussion (Grommers 1997, p. 204) the question was: "It is applicable at the level of the genome, the individual animal and even at the level of a breed or a species".

Taking for granted that our discussion here concerns the individual animal, then integrity shows another advantage. The focus remains upon animals' concerns and the likelihood of anthropomorphism is avoided. Further, integrity makes it possible to differentiate steps and grades of interference with an animal, "the wholeness and completeness, the balance in species specificity, and the capacity (for an animal) to independently maintain itself" (quoted from above) vary in degree, whereas dignity seems to be a concept of all or nothing. But a closer look reveals that the criteria mentioned above are not as easy to put in place as it may seem. For example, one must answer the question of what is an "appropriate environment" for an animal?

In case of farm animals, should this be the natural habitat of its wild ancestors? And what does species-specificity really mean when we are dealing with animals, where the varieties have undergone thousands of years of human breeding and selection?

Some, like recently Kerstin Schmidt (2008), have stated a clear-cut preference of the notion of integrity rather than adopting the principle of dignity. She quotes (2008, p. 215) Hauskeller with his phrase that "integrity is actually more concrete, more capable of factual description than the term dignity". One of the main arguments is that integrity can be classified according to some provable measure. Schmidt (2008, pp. 38–40) for example sets up a catalogue of four levels of interfering with animal integrity: ranging from the addition of new properties without reducing the original ones (like in the case of gene farming) as level 1 to level 4, the extreme of which is represented by the AMLs. Therefore, it should be possible to judge how relevant any interference with the inherited properties of an animal is.

7.5 Integrity and Dignity

But integrity in its pure sense is simply a way of labelling a certain state of affairs. Even Schmidt refers to an additional (in her case "zoocentric") ethical theory that frames her concept of dignity and gives it practical relevance. In other words, although integrity is a powerful tool to judge how intensively humans touch nature and the welfare of animals, dignity has a much broader function. When comparing the two terms it becomes clear that dignity serves a different purpose. Preserving integrity may be one criterion of treating living beings with due respect. Respecting the dignity of creatures demands this very kind of respect. It is simply not true, as Schmidt states, that the mere declamation of the dignity simply borrows its normative relevance from its similarity to human dignity, without giving proper reasons for this relevance. There have been fruitful attempts to explain what is meant by the term and why it is useful in debates on animal ethics.

Since the rather early expert opinion provided by Balzer, Rippe and Schaber (1998) on the article in the Swiss Constitution, one important concept has been used to explain how and why the dignity of creatures and especially of animals has to be respected: this concept was the concept of inherent value.

Verhoog and Visser (1997, p. 230) gave the following explanation. "In our view, therefore, the moral status of any organism is determined both by the fact that it is alive as well as by its specific nature. ... Being alive makes an animal relevant and this obliges moral agents to respect their intrinsic value. This obligation is made concrete by taking into account the specific nature of the organism concerned". This comes pretty close to one of my former attempts to give a more precise version of dignity and why we should keep in mind dignity of animals when dealing with them. In my book (Kunzmann 2007), I explained that dignity of animals has two dimensions with regard to animals, namely its inherent value (Eigenwert) and its proper nature (Eigenart). As far as I can judge, these two perspectives have reappeared in one version or the other. Once again Verhoog and Visser (1997, p. 229f):

Recognition of the intrinsic value of living entities is a necessary condition for attributing moral relevance to them, but not sufficient for assessing what exactly we owe to these entities, what kind of moral obligations we have to plants, animals or human beings. The nature of these specific obligations is determined by the species-specific characteristics which are essential for the living being concerned to be able to realize its good. Being alive always manifests itself in a (species-) characteristic way. The characteristics defining the specific nature of a living being constitute an important normative element when we have to determine what we owe to (what we ought to do or not to do in relation to) distinctive classes of living beings or to individual living beings in particular situations.

This gives us rather a complete program of adjusting the dignity of animals.

Let us begin with the so-called inherent value. Truly Balzer, Rippe and Schaber identified animal dignity (or to be more precise, the dignity of creatures) with their inherent value; this was a kind of reaction to the problem that they could not extend the core meaning of human dignity to other living creatures. They thought the core idea of dignity would be the "right not to be humiliated" (das Recht, nicht erniedrigt zu werden). This, in turn, presupposes a certain capacity of self-esteem which cannot be attributed to animals, the authors claimed.

So they focussed on "inherent value". But what does it mean to talk of an inherent value?

One aspect of the term is that animals should be treated well for their own sake. This is not only a prominent feature; it is the basis for modern animal ethics. The opposite would be to treat animals well, for this in the end serves human needs best. We try to preserve animal welfare because it is good for the animals in the first place.

7.6 Intrinsic Value and Dignity

What exactly can be called an "intrinsic value"? In the given formula, intrinsic value is just opposed to the so-called "instrumental value". This does not tell us much. There is a third possible version, which can be called an inherent value. Paul Taylor gave the following precise definition, as quoted in Heeger and Brom (2001, p. 243).

> What he [Taylor] proposes hinges on the concept of inherent worth. This concept ties together (1) the assertions about the good of a living being and (2) the demands of moral consideration for this being. With the concept of inherent worth, Taylor makes the claim that the good of a living being matters morally: We have direct duties towards this being... we ought to have moral consideration for it and promote the realization of its good.

The systematic gain of Taylor's approach is that it is not the living being in itself which can be considered as intrinsic value, but the states of affairs which are considered as good for the living being. As Rutgers and Heeger (1999, p. 43) put it, "By 'good of their own' (or 'natural good') we mean that animals have ends and purposes of themselves that are characteristics to them".

In a rather simple formula one could say: animals (and plants) are to be considered morally, because they have their own "meaning" (Sinn). It points to some kind of purpose and it makes sense for them to reach this purpose. Indeed, in current discussions a term for it became the Aristotelian (or Pseudo-Aristotelian) coinage of "telos". It is true that using the term telos can introduce myriad philosophical problems and telos can imply a wide variety of biological and bioethical meanings. It is not the purpose here to deal with all the implications. Rather, it should suffice to note that plants and animals (Kunzmann and Odparlik 2008) possess some capacity to cope with their environment and to stay alive and in good condition within given circumstances. To put it as bluntly as possible and to stay clear of so many theoretical quagmires: one can hardly deny that plants and animals prefer certain "states of affairs" compared to others. So it is not (only) the living being, say the animal, which in itself should be considered to be or to have an inherent value – it should be the state of affairs preferred by an animal, which has its inherent value. This can happen without any of the animals' actual feelings; as of course animals regenerate themselves constantly. When they are hungry they eat. When they are thirsty they drink. It does not matter, how intensely they "know" or they "feel" about the situation. Even plants can "decide" where the actual states of affairs are somewhat "desirable" for them. This is the point where talking of inherent value is useful. For living beings, some states of affairs are preferable in comparison to others and living beings incline towards those states of affairs that are closer to their needs and to their purposes. "According to Paul Taylor... each organism is a 'teleological (goal-oriented) center of life, pursuing its own good in its own unique way'. This is not to say that the organism's pursuit of its own good is necessarily conscious or intentional, but rather that 'a living thing is conceived as a unified system of organized activity, the constant tendency of which is to preserve its existence by protecting and promoting its well-being'" (Hauskeller 2005, p. 2).

"Helping" animals to achieve these goals is in the broadest sense caring for animal welfare. The interesting point here is that this new version combines the notion of intrinsic value with what we could call the "nature of an animal". For what may be called "the good" for an animal is prescribed by its "nature". In this dialectic structure, taking into account the inherent value (Eigenwert) is closely connected to taking into account the proper nature (Eigenart), for what is good for an animal is decided by its nature.

7.7 Species and *Eidos*

I am taking refuge in uncommon notions: The reason is that the usual ones carry to much historical or systematic bias. "Species" for example does matter for Martha Nussbaum's "capabilities approach" as in her book "Frontiers of Justice", where she speaks of "dignified life" (2006, p. 326), or of "animal dignity" (2006, p. 327) or of "dignity of a form of life" (2006, p. 346).

Even if we leave aside how problematic the term species has become in biology, its use in our context bears many possible misunderstandings, of which the most important (would be) to abandon the idea that species matters only inasmuch as the goals and aims of the individual are concerned. I would rather speak of eidos (which

7 Biotechnology, Battery Farming and Animal Dignity   111

of course bears many new risks). The argument is this: The "specific" properties of a certain type of animal are relevant only to the extent that they are properties of the individuals. One of the basic principles in animal welfare ethics is that welfare exists only in individual living beings (Nussbaum 2006, p. 358). Eidos is also chosen because it indicates that it is our "image" or "imagination" of an animal that rules our judgement on how we should treat it. Eidos may even be taken in a phenomenological sense – the essential point is that our treatment of animals should be in line with our imagination. Nussbaum uses the term "adequacy" for this idea.

I consider it to be a basic feature of all possible uses of the phrase dignity that we refer to an "adequate" behaviour towards someone or something. It is not the whole content of the word but an essential part of it. (In classical semantics one could speak of the ratio communis.) What do we mean when we refer to human dignity? We mean that something is appropriate (or in most cases: inappropriate) in connection to human beings, be it a certain behaviour, be it a certain set of circumstances. It is this common feature of "appropriateness" or "adequacy" that relates the uses of the phrase "dignity" when applied to humans and to animals. It does not mean of course, that we should treat them both in the same way. Quite on the contrary: we should treat animals as animals should be treated, more specifically as animals of a certain kind should be taken care of according to their "species-specific" needs and possibilities. By the way, it is therefore not necessary that treating animals with dignity excludes all those practices that respect for human dignity would forbid us.

To claim that dignity is an absolute concept, is not really an argument against the use of dignity in relation to animals. Human dignity may imply absolute claims. In our understanding, respect for human dignity excludes many of the things which we impose on animals. But this does not count as an argument, for this is in the strict sense human dignity. The history of the word is not necessarily confined to its use for humans only and therefore the features that we combine with human dignity are not strictly inherent to the term. As I said, a common feature is adequate behaviour. In the case of animals, according to Nussbaum's approach, we must be aware of their specific capacities. This does not necessarily exclude certain practices that are excluded with respect to human dignity, for example, keeping them in confinement or buying and selling them. Depriving animals of living out their "capabilities" is nonetheless ethically important and this is a basic result of discussing the dignity of animals. Animals have goods of their own according to their "eidos", and respecting animal dignity means at least trying not to deprive them of these goods – for the sake of the animals.

7.8 *Telos* ?

One may be inclined to quote this in the perspective of telos, as Rollin did: "It appeared to me (and still does) that the notion of telos provides a sound basis from which to deduce the rights of animals of different species, even as human rights have been based in plausible hypotheses about human nature. In so far as different kinds of animals are built in different sorts of ways for different sorts of things and in so far as these differences are rooted biologically, empirically ascertainable, environmentally expressed 'blueprints'" (in Ferrari 2008, p. 173).

Still, the concept of telos remains highly problematic. What Rollin means comes quite close to my understanding of animal dignity, however, especially with his focus on different kinds of animals. Acceptable treatment of animals is measured in the way it allows these blueprints to become reality. One has to keep in mind that this does not necessarily mean to ban all kinds of animal husbandry – it simply means we have to take it into consideration, and that we have to have good grounds for obstructing it. Nussbaum in turn judges an animal (Ferrari 2008, p. 159) "to be a complex living being with its intrinsic value; to respect this means to give it the opportunity to realize its specific properties or its unhindered flourishing".[12]

One has also to keep in mind that this does not exclude but includes animal suffering as a very important issue in keeping farm animals: As long as the animals are healthy and show no (physiological; behavioural) signs of distress, we may suppose that the impact on the animals is relatively low. But still, the introduction of perspectives like the "telos-approach" or the "flourishing-approach" are considerable steps beyond pain and injury.

7.9 Treating Animals Adequately

Are these approaches enough to judge the "change" of animals by breeding or biotechnology?

I do not think so.

What we could call the flourishing-approach simply takes place between the blueprint and the full realization of all the capacities that can be thought of within a given genetic setting. (In this line of thinking, many practices of modern battery farming need serious reconsideration.)

But what can we do, when the blueprint in itself is changed? Then, of course, the animal should be given different opportunities to realize different capabilities. If it were possible to breed birds without a nesting drive (Rollin 1995a, b), it simply would not be a breach of their dignity not to let them nest. If it is no longer part of their range of capabilities, would it be a good thing for them? Blind laying hens[13] would not feel the need to pick their neighbours' feathers.

---

[12] For her an animal is "eine komplexe Lebensform mit einem intrinsischen Wert, deren Achtung sich in der Ermöglichung zur Verwirklichung seiner spezifischen Merkmale bzw. seiner freien Entfaltung (flourishing) ausdrückt". (Ferrari 2008, p. 159).

[13] Rippe (2002, p. 240): "Es wird eine Hühnerrasse gezüchtet, in der alle Tiere blind geboren werden. Durch die Erblindung reduzieren sich Federpicken und Kannibalismus. Die Hühnerrasse zeichnet sich Umständen der gegenüber anderen extensiv genutzten Hühnerrassen also durch ein gesteigertes Wohlbefinden aus. Pathozentriker werden hier mindestens zugestehen müssen, dass im Vergleich zu anderen extensiv genutzten Hühnerrassen eine Verbesserung eingetreten ist".

Even in the bizarre case Douglas Adams imagined, namely a pig specially bred for being happy to be slaughtered – "a casserole of me perhaps?" – the principle would lead to accepting this strange behaviour as part of its nature.

This flatly contradicts our moral intuitions. We have to think of something else.

One way out was the introduction of genetic integrity. There are, however, serious concerns that make it inadvisable to follow that track. One is that integrity makes sense if and only if the "parts, the organs, the structures of an animal end up at what can really be called an 'integrum', a 'wholeness' and 'completeness'".[14] In the case of the genome, I cannot see this kind of wholeness and completeness.

The second objection is an ethical one. It makes sense to take care of animals because they know better and worse states, and they seek for themselves to reach the better ones or to persevere in them. All this gives rise to moral considerations. None of this can be said about genomes. In other words: talking about genomes does not generate a moral perspective. Genomes are apparently simply biochemical structures. Whatever they are, however, it is clear that they are nothing that in themselves can be the source of moral considerations.

They can, however, provide a cautionary tale. Changing single properties within an organism may collide with many other properties. These are extraordinarily complex entities. The integrity approach captures this nuance by accounting for the various stages of interference with animals and the various impacts it may have on the overall fitness and welfare of animals. The idea to breed blind laying hens for example must take into consideration that it is not the only function of the hen's body to be switched off; it is taking away a central organ of apprehension for a bird. Yet, one can still say that this has to be qualified from case to case. It is nonetheless plausible to think that extreme results in the breeding of farm animals may have considerable side-effects on their overall constitution.

Something else to be considered is "aesthetics". Whether we are considering battery farming or biotechnological concerns, many objections arise for aesthetic reasons. Think of caged laying hens or of broilers packed together. Think of naked cats. Think of gigantic salmon. Of course they make a strong impression. But is this all not only a question of getting accustomed to it? Think of common carp without scales – nobody is bothered, although it is quite natural (Kunzmann 2007, p. 125). Think of the enormous range of dogs – imagine, this variety would have been created by biotechnology and you can assume that fierce resistance would have followed.

Maybe.

Even here animal dignity offers some advice. The EKAH stated in 2000 that respecting the dignity of animals would exclude the following[15]:

> Interference with the appearance of animals

---

[14] Cf. Hauskeller (2005): "Now, for Aristotle, because every part of an organism's body is what it is in virtue of the whole whose part it is and for the sake of which it exists, it loses its identity when separated from that whole".

[15] "– Eingriff ins Erscheinungsbild – Erniedrigung – übermässige Instrumentalisierung". (See also Goetschel and Bolliger 2003, p. 239).

Submission humiliation
Reducing animals to mere instruments

Now we know that since the Balzer/Rippe/Schaber Report (1998, p. 28), the humiliation of a human person is the core idea of breaching human dignity which includes "das Recht, nicht erniedrigt zu werden". This in turn, makes no sense with animals that have no estimation of their own worth that could be offended. Why should it be morally wrong to change the appearance of animals, aesthetic considerations left aside? What does it mean to reduce animals to mere instruments (übermäßige Instrumentalisierung) and why should this be morally wrong? Philosophers know the conception in connection to Kant's formula of human dignity (and its long record in legal history) – but how can this be transformed to fit to animals?

The answer is that referring to animal dignity there are two different aspects which I earlier referred to as "actions and attitudes" (Handlung und Haltung) (Kunzmann 2007). What we have discussed so far has taken into consideration that treating animals with dignity is an aspect of actions. The three prohibitions of the EKAH mention actions of course, but their real interest is attitudes. To reduce animals to their mere function does not really qualify certain types of action, but is a statement about human motives or intentions. Complete humiliation of an animal does not necessarily harm it – but exposing animals to bizarre situations for example tells you something about the humans that enjoy it.

This explains why the phrase has become so widespread: It pinpoints a certain attitude towards animals that troubles many people. To produce blind hens because they may fit better into modern systems of husbandry is submitting the bird completely to human interests. To create mice that develop automatically the desired form of cancer reduces them to mere instruments.

This kind of attitude has possibly stimulated the successful introduction of our conception. The question above regarding the opposition to freely modifying animals to suit our needs is not so much a question of animal welfare; it is rather a question of attitudes and intentions. It is almost a symbolic act of "taking care" not to extend human domination over all and everything. To ascribe dignity to animals is a very humane performance because it means that humans are capable of denying or at least of limiting their own interests. Some 25 years ago, R. Spaemann argued that it is a performance of which only human beings are capable, and which is at the basis of their own dignity. It belongs to the gifts and to the obligations of humans to let "the things be". Humans can limit their own will to dominate anything else and simply let other things have their own way. According to Spaemann, it is a true sign of human dignity to care for the nature of all the creatures of our local environment.[16]

---

[16] In Spaemann's (1984) words, it is the "Fähigkeit, der naturwüchsigen Expansion des eigenen Machtwillens Grenzen zu setzen, einen nicht auf eigene Bedürfnisse bezogenen Wert anzuerkennen, in der Fähigkeit, anderes in Freiheit 'sein zu lassen'... Es [macht] gerade die Menschenwürde aus, im Umgang mit der Wirklichkeit deren eigenen Wesen Rechnung zu tragen".

This final meaning of animal dignity addresses all our problematic cases and it is finally the reason why dignity is the more demanding concept. It includes not only animal welfare but also the human capacity for taking care. As with human dignity, it is not at all easy to give a positive description, but we can name those situations where people breach this dignity: In those cases when their own interests seem to override any consideration of the animals with the result that animals become instruments for production units, for sports equipment, for clowns, for measuring instruments, for lacking family members... For all these cases, the term dignity of animals could apply because we should treat animals in any circumstance adequately. This means that we should treat animals essentially and principally as animals.

References

Balzer P, Rippe K-P, Schaber P (1998) Menschenwürde vs. Würde der Kreatur. Alber, Freiburg, München
Baranzke H (2002) Würde der Kreatur. Königshausen & Neumann, Würzburg
Bressler H-P (1997) Ethische Probleme der Mensch-Tier-Beziehung. Eine Untersuchung philosophischer Positionen des 20. Jahrhunderts zum Tierschutz. Lang, Frankfurt am Main
Brom F W A (2000) The good life of creatures with dignity. J Agric Environ Ethics 13:53–63
EKAH (2000) Stellungnahme zur französischen. Version des Art. 120 BV. EKAH, Bern
EKAH/EKTV (2001) Würde des Tieres. Bern
Ferrari A (2008) Genmaus und Co. Harald Fischer, Erlangen
Goetschel A, Bolliger G (2003) Das Tier im Recht. Orell Füssli, Zürich
Grommers F J (1997) Consciousness, science and conscience. In: Dol M, et al. (eds) Animal consciousness and animal ethics. Van Gorkum, Assen
Hauskeller M (2005) Telos: The revival of an Aristotelian concept in present day ethics. Inquiry (Oslo) 48(1):62–75; available in PMC 2006 January 25
Heeger R (2000) Genetic engineering and the dignity of creatures. J Agric Environ Ethics 13:43–51
Heeger R, Brom F W A (2001) Intrinsic value and direct duties: From animal ethics towards environmental ethics? J Agric Environ Ethics 14:241–252
Kunzmann P (2005) Rinder ohne Hörner? Zwischen ethischen und wirtschaftlichen Überlegungen. BLW 31/2005
Kunzmann P (2007) Die Würde des Tieres zwischen Leerformel und Prinzip. Alber, Freiburg Br., München
Kunzmann P, Odparlik S (2008) Tiere und Pflanzen als Gegenstand der Forschung – eingedenk ihrer Würde. Bulletin der Vereinigung der Schweizerischen Hochschuldozierenden (VSH) 34. Jg., Nr 3/4, Nov 2008
Mepham B (2000) "Würde der Kreatur" and the common morality. J Agric Environ Ethics 13: 65–78
Löwer W (2001) Tierschutz als Staatsziel – Rechtliche Aspekte. In: Thiele F (ed.) Tierschutz als Staatsziel? Bad Neuenahr-Ahrweiler 25:31–49
Nussbaum M (2006) Frontiers of justice. Harvard University Press, Cambridge, MA, London
von der Pfordten D (2003) Tierwürde nach Analogie der Menschenwürde. In: Brenner A (ed) Tiere beschreiben. Harald Fischer, Erlangen, pp. 105–123
Richter D (2005) Sprachenordnung und Minderheitenschutz im Schweizerischen Bundesstaat. Springer, Berlin, Heidelberg, New York
Rippe KP (2002) Schadet es Kühen, Tiermehl zu fressen? In: Liechti M (ed) Die Würde des Tieres. Harald Fischer, Erlangen, pp. 233–242
Rollin B (1995a) Farm animal welfare: Social, bioethical and research issues. Iowa State University Press, Ames

Rollin BE (1995b) The Frankenstein syndrome: Ethical and social issues in the genetic engineering of animals. Cambridge University Press, Cambridge

Rutgers B, Heeger R (1999) Inherent worth and respect for animal integrity. In: Dol M, et al. (eds) Recognizing the intrinsic value of animals. Van Gorkum, Assen

Schmidt K (2008) Tierethische Probleme der Gentechnik. Mentis, Paderborn

Schneider M Die Würde der Kreatur. http://www.schweisfurth-stiftung.de/347.html

Schneider M (2001) Über die Würde des Tieres. Zur Ethik der Mensch-Tier-Beziehung. In: Schneider M (ed) Den Tieren gerecht werden. Zur Ethik und Kultur der Mensch-Tier-Beziehung. Universität Gesamthochschule Kassel, Witzenhausen

Spaemann R (1984) Tierschutz und Menschenwürde. In: Händel UM (ed) Tierschutz. Fischer, Frankfurt/Main

Steiger A (2002) Die Würde des Nutztieres – Nutztierhaltung zwischen Ethik und Profit. In: Liechti M (ed) Die Würde des Tieres. Harald Fischer, Erlangen, pp. 221–232

Verhoog H, Visser T (1997) A view of intrinsic value not based on animal consciousness. In: Dol M, et al. (eds) Animal consciousness and animal ethics. Van Gorkum, Assen

Part III
Food, Globalization, and Water

Chapter 8
Agricultural Trade and the Human Right to Food: The Case of Small Rice Producers in Ghana, Honduras, and Indonesia

Armin Paasch, Frank Garbers, and Thomas Hirsch

Abstract Rice production and rice policies have an immediate relevance for food security in the world. This article summarizes a comprehensive study commissioned by the Ecumenical Advocacy Alliance (EAA), and conducted by the FoodFirst Information and Action Network (FIAN) (see Paasch, Garbers and Hirsch 2007). The study examines the impact of specific rice trade policies on the Human Right to Adequate Food of specific rice producing communities in Ghana, Honduras, and Indonesia and analyzes causal chains, first between sharp increases of rice imports and hunger, malnutrition and food insecurity, and secondly between these import increases and certain trade and agricultural policies. The case studies also include a thorough human rights analysis of the findings, distinguishing between the different obligations and responsibilities of national governments, external States, and intergovernmental organizations (IGOs). We obtain insights into the causes for the replacement of local rice production in developing countries with imports and explanations why the populations of these countries have been hit so hard by the price increases.

8.1 The Right to Food in Times of Globalization

Access to adequate food is a basic human right for every person. It is enshrined in article 25 of the General Declaration of Human Rights and article 11 of the International Covenant on Economic, Social, and Cultural Rights (ICESCR) (UN 1976). The right to food, according to the authoritative interpretation of the UN Committee on Economic, Social, and Cultural Rights (CESCR), is not to be interpreted in the narrow sense of being fed, but rather means access at all times – physical and economic – to "adequate food" and the ability to procure it. Food must

A. Paasch (✉)
Misereor, Aachen, Germany
e-mail: armin.paasch@misereor.de

be adequate in terms of quantity and quality, as well as being culturally acceptable. And the enjoyment of the right to food must not threaten the "attainment and satisfaction of other basic needs" such as health, housing, and education (UN 1999).

160 States have ratified the ICESCR and are obliged to respect, protect, and fulfill the right to adequate food. Each State party has the obligation to develop strategies to progressively realize the right to food for all people, by using the "maximum of its available resources." Such strategies must address all aspects of the food system, including the production, processing, distribution, marketing, and consumption of food. Access to productive resources is a key element of the right to adequate food, especially in rural areas where almost 80% of hungry people live. Furthermore, people must be able to feed themselves in dignity from agricultural activities. Fair market conditions are a key part of an enabling environment which States are obliged to create in order to implement the right to adequate food.

States' obligations do not only refer to the people within the respective state's national borders, but also have an international dimension. Brot für die Welt, the German Church Development Service (EED), and the FoodFirst Information and Action Network (FIAN) have proposed the term "extraterritorial obligations" (ETO) to describe the international dimension of states' obligations, which are part of the ICESCR. This international dimension applies for the same levels of obligations as within national borders, but especially for the "minimum obligation" to respect, which, according to human rights experts, is already part of existing human rights legislation. Hence, no State shall do harm to the right to adequate food of people living in other countries (Windfuhr 2005).

This obligation is especially relevant when it comes to development aid, international investment, or trade. Dumping or forced market opening, when they lead to the destruction of local market access, the income basis and food security of peasants, are possible examples of extraterritorial violations of the right to food. The obligation to respect the right to food abroad does not only refer to bilateral relations but also includes decisions within international organizations such as the World Bank, International Monetary Fund (IMF), or World Trade Organization (WTO). According to the CESCR, "states should, in international agreements whenever relevant, ensure that the right to adequate food is given due attention and consider the necessary development of further international legal instruments to this end."

8.2 Rice Trade Liberalization as a Threat to Small Producers

Rice was chosen as an example for this study, because it is central for food security all over the world. It is the main source of calories for half of the world's population and the main source of income and employment for two billion people, most of them peasants, and most of them women (FAO 2004). Only 6.5% of the global rice consumption is traded internationally, the biggest exporting countries currently being

Thailand, Vietnam, India, the US, and Pakistan (World Bank 2005). Nevertheless, international rice trade can have a serious impact on the development of national rice markets and prices. Following the price explosion for food staples, food riots have taken place in more than 30 countries since the beginning of the year 2008 and reminded us the close links between international markets and food security. Rice is the commodity most affected by the price explosion. In many countries, such as Haiti and Honduras, the increase in rice prices is one of the main reasons for these riots. While soaring commodity prices are the cause of the current crisis, we should not forget that this price explosion had been preceded by a long period of price decline since the 1970s. While many countries now have difficulties to purchase rice at the international markets, for a long time, the main problem had been the opposite in many countries. The Food and Agriculture Organization of the United Nations (FAO) has registered 408 cases of import surges for rice in 102 countries between 1983 and 2003, with a disquieting concentration of them in Africa, the Pacific Islands, and Central America (FAO 2007a).

In many countries, these imports have replaced domestic production to a large extend; they have made these countries reliant on imports and, as a result, very vulnerable for price fluctuations in the international markets. The current "food crisis" has not come over night.

Among the complex factors, three policy reasons can be identified as endemic and appear most frequently to boost import surges, and, more generally, import increases: (1) Markets have been opened to imports in many countries since the beginning of the 1980s as a result of Structural Adjustment Programmes (SAPs) often imposed by the IMF and the World Bank, regional free trade agreements and, to a lesser degree, the Agreement on Agriculture (AoA) of the WTO. (2) The high levels of support for production, processing, and export of rice in some developed countries have contributed to import surges, which have occurred most frequently in times of very low world market prices for rice, such as in the years 2000–2003. According to Oxfam, the US exported rice at 34% below production costs in 2003, a practice which can be described as dumping (Oxfam 2005). (3) The cutting of support for agricultural inputs, machinery, public procurement and price guarantees etc. in many developing countries, as part of the same SAP mentioned above, has often resulted in the reduction or stagnation of domestic rice production capacities. Instead of supporting those capacities, many governments prefer to fill the gap in supply with cheap imports.

The case studies summarized in this article include an overview of the development of rice imports and domestic rice production at a macro-level, and an analysis of the domestic rice policies including border measures. They also include an analysis of possible dumping practices by countries of origin of rice imports and possible pressure that other countries may have exerted on Honduras, Ghana, and Indonesia, through bilateral or multilateral trade agreements or IGOs, to adopt certain rice trade policies. And they include, as a core component, qualitative analysis of the possible impact of increased rice imports on the incomes, livelihoods, and food security in selected rice-producing communities. Finally, the studies conclude with an analysis of states' behavior from the perspective of the human right to food.

The main challenge of the studies is the verification of possible causal links first between sharp increases of rice imports and hunger or malnutrition in the communities, and second between high imports and certain trade and agricultural policies. This verification of causalities up to a violation of the right to adequate food requires a careful assessment of other additional factors which might have worsened access to food for the rice farmers, such as natural disasters, violent conflicts or wars, possible changes in land tenure arrangements or deteriorated access to infrastructure, farm inputs, credits or extensions services. Another challenge for the human rights analysis is to distinguish between the responsibilities of different states for these trade policies. In many cases, national governments, IGOs and other external state actors share the responsibility. Only if such causality can be verified and clear state responsibilities can be identified, is it possible to document a violation of the right to adequate food.

8.3 Ghana: Rice Liberalization Under the Auspices of the IMF

Whereas until recently, rice had been a niche product for urban elites, demand has grown remarkably over the last 10 years in Ghana. This development could have opened a window of opportunity for growth in domestic rice production and reduction of poverty among the estimated 800,000 Ghanaian rice producers, who are predominantly smallholders. However, the opposite is the case. From 1998 to 2005, the area planted with rice diminished from 130,000 to 120,000 ha and the annual paddy production level from 281,000 to 237,000 Metric tons (Mt) (MOFA 2006). Studies have indicated that incomes of the farmers have been declining over recent years, with alarming effects in terms of poverty and food insecurity. This crisis hits a part of the population that is highly affected by poverty and vulnerable to hunger. The incidence of poverty is almost 60% among food crop farmers, 70% of them being women.

The explanation for this paradoxical development is that the growing demand for rice in Ghana has been captured entirely by imports, mainly coming from the US, Vietnam, and Thailand. From 1998 to 2003, imports rose from 250,000 Mt to 415,000 Mt, an increase of nearly 70%. The market share of local rice fell from 43% in 2000 to only 29% in 2003. An "import surge," according to the FAO, occurred especially between 2002 and 2003, when the volume of rice imports grew by 154%, while the volume in domestic rice paddy production declined from 280,000 to 239,000 Mt, representing a 16% decline (Asuming-Brempong 2006 and FAO 2006a).

The main reasons for the import surges seem to be the low level of world market prices for rice between 2000 and 2003 and the low CIF prices of the imported rice, which on average have consistently been below the wholesale price of domestic rice in Ghana (Action Aid 2005). While local rice with poor processing quality is often sold more cheaply in the market, imported rice generally beats the local kind because of its better price–quality relation. High-quality imported rice is generally cheaper than high-quality local rice. Another advantage of imported rice lies in better marketing channels of the highly concentrated rice import business, which make it available everywhere, while local rice can hardly be found in some urban markets and almost completely fails to reach big consumers such as restaurants and hotels any more.

The study shows strong evidence that three policy reasons have contributed to the boost of imports:

(1) The removal of import controls and the introduction of a low tariff applied on rice imports of 20% in 1992 led to import increases over the 1990s (Assuming-Brempong et al. 2006). The attempt of the Ghanaian government and parliament in 2003 to increase the rice tariff from 20 to 25% (and the tariff for chicken from 20 to 40%) through Act 641 to respond to increased imports was obstructed by external actors, especially the IMF. The implementation was suspended only 4 days after it had started. The IMF report on the Article IV Consultations on the Ghana Poverty Reduction Strategy reveals that, during these consultations, the authorities committed to tariff increases not being implemented "during the period of the proposed arrangement" (IMF 2003a). Indeed, on May 9, the Executive Board of the IMF concluded these Article IV Consultations and approved a 3-year arrangement under the Poverty Reduction and Growth Facility amounting to SDR 185.5 million (258 million USD) and additional interim assistance under the Initiative for Highly Indebted Poor Countries of SDR 15.15 million (about 22 million USD) (IMF 2003b). And on May 12, just 3 days after the approval of the IMF loan, the suspension of the implementation of Act 641 was issued. Thus, the same consultations that led to the approval of the loan also "convinced" the Ghanaian government to cut the tariffs back to the previous level.

(2) The second policy reason is the high margins of dumping for rice imported from the US, Vietnam, and Thailand. According to calculations commissioned by Oxfam on the three main countries of origin for 2003, the export prices were far below the home market prices ("normal values") of selected rice varieties imported to Ghana. For the US rice varieties, the highest margins of dumping were found on average (Ayine 2006). US dumping was also evident while comparing export prices with production, the former being 34% below the latter in the year 2003. Dumping is an important reason for the fact that, in terms of prices, imported rice can compete with and often is even cheaper than Ghanaian rice (Oxfam 2005).

(3) The progressive removal of support to the Ghanaian rice sector between 1983 and the late 1990s resulted in an extremely poor national infrastructure for the production, processing, and marketing of rice, leading to serious supply constraints of the domestic rice sector in terms of quantity and quality (JICA 2006). It removed support, which formerly facilitated access to credits, seeds, fertilizers, the use of machinery at favorable conditions and marketing. These policies, to a large extent, followed the SAPs introduced by the IMF and the World Bank since 1983 (Khor and Hormeku 2006).

The micro-level study on the effects of imports was conducted in Dalun, a rice-farming community with 10,000 inhabitants in the Tolon Kumbungu District of the Northern Region, located at about 50 km from the region's capital Tamale. All the market women interviewed stated that, especially since 2000, imported rice has taken over the Tamale market to a large extent. As a result, the quantity of rice bought by Tamale market women in Dalun and the surrounding villages and sold in Tamale has diminished dramatically by around 75%. This information coincides with that provided by the local miller, whose processing volume declined in a similar dimension, and the farmers' experience that they are selling much less paddy than before. Farmers additionally suffered a dramatic decline in real producer prices since 2000. While from June 2000 to 2003, the Ghanaian currency Cedi lost 46% of its value, the nominal prices fell considerably according to some farmers and remained stagnant according to others. In both cases the drop in real prices is dramatic. This had a direct negative impact on the incomes of farmers because the real costs of production only decreased moderately at the same time.

As a result, rice farming families are increasingly suffering malnutrition and food insecurity. All the interviewed peasants report that their families are suffering hunger. They do not have stable access to adequate food because, in the period before harvest, most of them have to reduce meals in number, size, and quality. Health problems among the children who are most affected by this shortage of food are mentioned frequently in the interviews. And the incomes of peasants have declined in a way that they are burdened with debt and lack money reserves. In case of a loss of yield due to unexpected shocks such as droughts or pests, the health of peasant families and especially of the children would be heavily affected. In addition, peasants report that they have to spend a larger share of their income to purchase food and, especially in the same "period of hunger," have to reduce their expenditures required to enjoy other basic human rights such as the rights to health and education.

To conclude, there is strong evidence that a combination of import liberalization, dumping, and the removal of domestic state support has significantly increased malnutrition and food insecurity and thus led to a violation of the human right to adequate food of peasant families in Dalun. Three actors are mainly responsible for these policies and have breached their obligations and/or responsibilities under the right to food: (1) the state of Ghana breached its obligation to protect the right to adequate food of rice peasants in Dalun and elsewhere by cutting market protection in 1992 and by not increasing protection later despite the evident injuries caused by imports. By cutting existing support to rice farmers, Ghana also breached its obligation to respect the right to food of the peasants, and its obligation to fulfill the right to food by applying policies that do not create an enabling environment of these families to feed themselves. (2) The IMF breached its responsibility to respect the right to adequate food by pressuring the Ghanaian government to remove support and protection for poor rice peasants over the 1980s and 1990s and by pressuring the government to suspend Act 641 in 2003. And consequently, the member states of the IMF thereby breached their obligation to respect the right to food of the rice peasant families in Dalun and elsewhere. (3) And finally, the exporting rice countries involved in dumping practices, especially the US, have breached their obligation to respect the right to food of rice peasant families. Its subsidies, export credits, and the misuse of food aid have contributed to the displacement of domestic rice from the markets of cities like Tamale and to the losses of income of the rice peasants in Dalun.

8.4 Honduras: Natural and Manmade Disasters

In Honduras, half of the nearly 8 million inhabitants live in rural areas and are either directly or indirectly linked to agricultural production. Around 70% of rural households live under the poverty line (UNDP 2006). Rice represents the third most important staple food in Honduras after maize and beans. Average per capita consumption of rice per year increased from 8 kg in 1993 to 16 kg in 2004 (Ponce Sauceda 2004). Paradoxically, in the same period, rice production suffered an unprecedented fall. While the annual paddy production had increased between 1966 and 1990 from 9,300 to 47,300 Mt, during the 1990s it decreased dramatically down to a level of 7,200 Mt in 2000. The rice market was, to a large extent, taken over by imports coming from the US (FAOSTAT).

The transformation of the rice sector, which took place in less than 15 years, can be divided and analyzed in three stages:

(1) The liberalization of agricultural markets started in 1991, when the Honduran Institute of Agricultural Marketing (IHMA) disappeared and guaranteed prices were withdrawn. After the submission of the executive decree to eliminate the State's monopoly on the foreign trade of agricultural products, the Congress of the Republic took advantage of a period of legal vacuum to temporarily allow the import of rice at a reduced tariff of 1%. Imports shot up immediately. In a good rice harvest year of over 54,000 Mt (paddy), 32,000 Mt of milled rice and 12,500 Mt of paddy rice were imported. The market practically collapsed. The FAO describes this sudden liberalization of the market – known as the *arrozazo* – as an import surge. Farm gate prices fell by 13% in 1991 and by 30% in 1992. In 1992, a price-band mechanism was introduced, which allowed an increase of tariffs up to 45% depending on the price in the international market. Imports decreased in 1992 and 1993 to the levels prior to the *arrozazo*. However, a new period of massive imports started in 1996, this time rather focused on milled rice. New phytosanitary regulations for paddy rice imports and the unification of price-band mechanisms for milled and paddy rice (pleading the same tariff level for both products) had boosted milled rice imports. As a consequence, the milling industry was neglected, and its market position was negatively affected. The milling industry bought less Honduran paddy rice, which led to the reduction of the cultivated surface from 16 to 10,000 ha between 1997 and 1998. The negative impact of imports was worsened through the liberalization of the market for agricultural inputs, credits, and land, as a

result of the Law for the Modernization of the Agricultural Sector (LMA) of 1992. Both the reduction of import protection and of producer support had been part of SAPs introduced in Honduras following the advice of the IMF and the World Bank (Oxfam 2004, World Bank 1994, FAO 2007b).

(2) The rice sector was already in a deep crisis when it was hit by Hurricane Mitch in 1998 and later in 2001 by tropical storm Michelle. Entire regions, like the south of Honduras, practically disappeared from the national rice production map. There is strong evidence that the devastating effects of natural disasters on rice farmers were worsened through disaster relief measures. It is amazing that, even in the 2 years following Hurricane Mitch, when rice production was considerably reduced, farm gate prices did not increase. The reason was the high rice supply at a low price as a result of dumped imports (Oxfam 2004). From 1999 onwards, the price of US imported paddy rice in Honduras fell below the price received by Honduran producers. Apart from commercial exports, food aid – 70% of it from the US – contributed to a rice surplus in the Honduran market after Hurricane Mitch. Summing up national production, commercial imports and food aid, in 1999 there was an additional rice supply compared to 1998 of 34,000 Mt of milled rice and 25,000 Mt of paddy rice.[1] There is no doubt that food aid was necessary after Hurricane Mitch in order to guarantee food security. However, the amount and prolonged period of food aid turned into an instrument of dumping and caused adverse effects. Under these conditions, the rice sector had almost no possibility to recover. Thousands of rice producers who could overcome the decade of the 1990s could not avoid financial collapse at the beginning of the new millennium.

(3) As a response to the long-term crisis of the rice sector, the Rice Agreement between the rice industry (national rice millers), the productive sector, and the government was signed in 1999. The Agreement, which is still in place, states that millers can import paddy rice with a preferential tariff of 1%, as long as they buy practically all the national production at the price stipulated by the parties annually. At the same time, the tariff for milled rice and other types of rice is kept at 45%. On the one hand, the agreement allows 22 national millers to establish an oligopoly over the national rice supply, be it national or imported rice, and to dominate the whole rice chain from production to retailing in Honduras. On the other hand, it clearly had a stabilizing effect on the rice productive sector. Those peasants who are part of the agreement experienced a certain recovery on a low level and reached some income security. However, the Free Trade Agreement between the US and Central American countries plus the Dominican Republic (DR-CAFTA), which has been in place since April 2006, will undermine these achievements and probably cause a slow death for rice production in Honduras. Once this 10 year, 45% maximum tariff period is over, the tariff on rice imports will be progressively reduced to 0% within 8 years (until 2024), and Honduran rice producers will be totally exposed to dumped

---

[1] Calculation made by author using data from FAO, USDA, and SIECA.

rice imports from the US. The US refused to talk about its own subsidies to the rice sector in the negotiations on DR-CAFTA (Ponce Sauceda 2003, Garbers and Gauster 2004).

The micro-level study on the impact of imports was carried out in two rice-producing communities: the Guangolola community in the Yoro Department, and the Guayamán community, in the Otoro valley, in the Intibucá Department. Both communities are organized in associative businesses or cooperatives, the Cooperativa Agropecuaria Regional El Negrito Limitada (CARNEL) in Guangolola, and the Empresa Asociativa de Campesinos de Transformación y Servicios Otoreña (EACTSO) in Guayamán. Both communities had been beneficiaries of the Agrarian Reform, and had overcome the level of subsistence farming thanks to the dynamic development of the rice sector during the 1980s in Honduras.

Guangolola producers remember the *arrozazo* as the end of a period of growth and the beginning of the difficult decade of the 1990s, as the mills refused to accept their produce and the producers got into debt. In 1998, Hurricane Mitch destroyed the productive infrastructure, including the whole community irrigation system. Three years later, Hurricane Michelle had the same effect. The rice food aid given between these two disasters additionally affected the marketing of the restored production. The key factor to restoring rice production after 2001 was the Rice Agreement. Rice producers in Guangolola achieved a level of economic stability that they had not had since the early 1990s.

In Guayamán too, farmers remember the *arrozazo* as the beginning of a long crisis. Additionally, the reduction of support and the consequent price rise for inputs are reported to be a key factor for the decline. The destruction caused by the natural disasters was aggravated by the fact that food aid saturated the market afterwards. Today, out of 30 families, only four or five maintain individual rice production and cultivate an area of around 20 *manzanas*. They channel their production directly through the millers, as established in the Rice Agreement, as the mill cartel did not accept their cooperative EACTSO as a processing business in the framework of the Rice Agreement. Most of the community members try to ensure their livelihood by complementing their incomes from grain production by paid jobs in the region or in Tegucigalpa, although with limited success.

As testimonies show, the economic security level both in Guayamán and Guangolola has been significantly reduced since the beginning of the 1990s. Particularly in the case of Guayamán, there is evidence that, at least during certain phases of the crisis, rice-producing families suffered a reduction of the quantity and quality of the food available. The achievements of the Agrarian Reform have suffered a roll back in both communities as land concentration has increased again. In Guayamán, producers returned permanently, and Guangolola producers temporarily, to subsistence agriculture, a scheme they had overcome at the beginning of the 1990s. Many peasants are burdened with debt and the rice crisis blocks their future development potential. In both communities, vulnerability in terms of food security in case of external shocks is extremely high.

To conclude, the right to adequate food of the rice farming families in Guayamán and Guangolola has been violated through agriculture and trade policies.

(1) The state of Honduras breached its obligation to respect the right to food by cutting support to the peasants through the LMA in 1992 and increasing input costs through the devaluation of the Lempira. Honduras breached the obligation to respect the right to food of the Guayamán families by cutting their market access through restrictive affiliation rules of the Rice Agreement, without providing any alternative. Honduras breached its obligation to protect the right to food by opening the market to dumped imports in 1991 and later in 1996, despite the obvious injury caused to poor peasants. Through deficient crisis management after Hurricane Mitch, it allowed extensive food aid to come in and contributed to deepening the crisis among these peasants. By ratifying the DR-CAFTA, the Honduran State has renounced the policy space, which is necessary for it to protect the right to food of domestic rice producers. And finally, through the implementation of DR-CAFTA, and earlier through the restrictions to development potential inherent to the Rice Agreement, the state of Honduras has failed to create an enabling environment for the realization of the right to adequate food of rice producers.
(2) The IMF and World Bank clearly breached their responsibility to respect the right to food in Honduras by forcing market opening and cuts in the support for poor rice producers. By supporting those policies, member states of the IMF and World Bank did not comply with their obligations to respect the right to adequate food in the rice-farming communities in Honduras.
(3) The US breached its obligation to respect the right to adequate food through dumped rice exports to Honduras from 1991 onwards and the misuse of food aid as an instrument of dumping. The US also breached its obligation to respect the right to adequate food by pushing for and signing DR-CAFTA, which threatens to displace poor Honduran rice producers from the market in increasing numbers.

8.5 Indonesia: The World's Third Biggest Rice Market Under Threat

With an annual production of 54.8 million Metric tons (2006), Indonesia is the world's third biggest paddy producer. Rice is produced by approximately 13.6 million farmers, of whom 65% are considered as poor smallholders with less than 0.5 ha of landholding size. It is estimated that about 21 million people find employment in the whole rice sector. At the same time, rice is by far the most important staple food for almost 215 million people, contributing 60% to the per capita daily calorie intake (FAO 2006b). Rice consumption currently exceeds production by approximately 5% and is developing even faster than production, also making Indonesia one of the world's largest rice importers.

8 Agricultural Trade and the Human Right to Food 129

Since independence in 1945, every government has wanted to achieve, in parallel, low rice prices for consumers and high income for paddy farmers. Indonesian's rice policies since 1967 can be structured into three phases:

(1) During 1967–1996, the government controlled the domestic rice market by intervening in the market in order to encourage production and to maintain price stability. The intervention took place in the form of managing huge governmental stocks via BULOG, the state-owned Logistic Affairs Agency (Badan Urusan Logistik). Imports were strictly regulated by tariff and import control policies and aimed at closing the gap between national production and consumption. Indonesia met its self-sufficiency target in 1984 and became a net rice exporter during 1985 and 1987. Since then the country has again become a net rice importer (Mulyo 2004).
(2) Import liberalization started as early as 1995 as a result of the AoA of the WTO. However, radical liberalization only occurred in 1997 under the pressures caused by the Asian economic crash. The Indonesian government was obliged to sign a Letter of Intent under the directive of the International Monetary Fund (IMF) including the commitment to implement SAPs. BULOG was privatized, and rice market support was stopped. The import tariff was nil and unlimited imports were allowed between 1998 and 1999. The government significantly reduced subsidies, including the agricultural input subsidies, which were highly relevant before. As a result of the new policy, imports boosted to six million Metric tons in 1998 – turning Indonesia into the world's biggest rice importer in that year – and four million Metric tons in 1999, mainly originating from Thailand, followed by Vietnam. By applying export credits and subsidizing agricultural inputs, both countries kept export prices artificially low and flooded the Indonesian market with dumped rice. Another reason for the crisis was a drought provoked by the southeastern branch of the climatic phenomenon El Niño in 1998. The loss in production through El Niño was only about 4–5%, while imports reached a 12% market share and over-compensated for the loss in production by far. During this time, Indonesia's self-sufficiency ratio decreased and the dependence on imports increased. Paddy farmers were severely hit by the disastrous combination of decreased paddy prices, higher prices for (less subsidized) agricultural inputs and the loss in production through El Niño (UNDP 2005). Because of the oligopolistic structure of the Indonesian rice market, liberalization did not even lead to lower prices for urban consumers. On the contrary, consumer prices increased in the period of liberalization.
(3) Because of the negative effects of market liberalization on producer and consumer prices the government has step by step returned to controlling the domestic rice market since 2001, with various modifications up to an import ban in 2004. The former policy of floor prices for unhusked paddy was replaced by the Government Procurement Price, which sets a ceiling price that has not been as effective. Only during periods of price hikes would BULOG perform market operations. Current trade policies particularly aim at stabilizing the

domestic price for unhusked paddy by a seasonal import ban and the management of rice stocks through the privatized BULOG (Nizwar et al. 2007). While the government has taken action in favor of the marginalized and extremely vulnerable paddy farmer, the system has not fully led to the expected results. Consumers are suffering from a price hike whereas producer prices remain comparatively low. It is the oligopoly of traders, which benefits most. Domestic agricultural support granted to producers is still a decisive factor for farmers' welfare. Despite increases in subsidies since 2003, the necessary level has not yet been reached, and support is not always well targeted to the farmers who most need it. Currently the World Bank is pushing for a cancellation of the import ban, granting import licenses to the private sector and tariffs of only 10–15%. The experience of the liberalization period indicates that such policies would threaten the livelihoods of millions of paddy farmers in Indonesia (UNDP 2005).

Field research on the effects of trade and rice policies was conducted in four communities – Cikuntul, Cikalong, Pinangsari, and Samudrajaya – in the three regions Subang, Karawang, and Bekasi of the province of Jawa Barat (West Java). The regions were selected since they form the center of rice production, more popularly known as West Java's rice barn. In all these communities, imports are reported to have a negative impact on the incomes of the farmers. Middlemen react to the increased supply through imports by lowering the prices offered to farmers. After the re-regulation of imports, middlemen managed to keep farm gate prices low, often misusing the lack of information of the farmers. Low prices have a direct negative impact on the incomes of the farmers since, at the same time, production costs and living costs are increasing.

For more than three decades, but in particular since rice market liberalization in 1997, the living conditions of the farmers have gotten worse, according to their own perception. All statistical data as well as interviews clearly indicate that most of the peasant families are living on less than one USD per person and day and that they cannot fulfill their basic needs regularly. Reducing food is a last resort amongst these communities. Most of the peasants feed themselves sufficiently in quantity but the diet is not always nutritionally appropriate. In order to buy enough food most of them cannot regularly meet their other basic needs like housing, health, and education. There is a considerable group of people who have to reduce their daily meals in the months before harvest due to lacking income.

It can be concluded that trade and agricultural policies have significantly contributed to violations of the right to adequate food of these rice-farming communities. They reinforce a multi-complex setting of negative framework conditions for paddy farming communities, such as inadequate access to land and other productive resources, inappropriate knowledge of the market, and high dependence on middlemen.

(1) The State of Indonesia breached its obligations to respect and to fulfill the right to food by severely reducing or even abolishing the domestic agricultural support in 1998 that had been granted to paddy farmers for decades, and

by abolishing substantial parts of governmental procurement prices and other price-stabilizing measures. It thereby also breached its obligation to fulfill the right to food by not creating an enabling environment for the realization of farming families. By opening the market for free rice trade in Indonesia in 1997/1998, the government has breached its obligation to protect paddy farmers' right to food. This led to a loss in market share and revenues so that many farmers fell below the poverty line and became unable to feed themselves adequately.

(2) By forcing the Indonesian government to implement these liberalization policies that increased food insecurity among rice peasants in the four communities, the IMF has breached its responsibility to respect the right to food. Similarly, by pushing for these measures in 1997/1998 and for similar policies currently, the World Bank is breaching its responsibility to respect the right to food of the peasants. The member states of IMF and the World Bank, accordingly, are breaching their obligation to respect the right to adequate food.

8.6 Conclusions of the Case Studies

The case studies show strong evidence that trade and agricultural liberalization has significantly contributed to the violation of the human right to adequate food of rice-farming communities in Ghana, Honduras, and Indonesia. Increased and cheap imports substantially reduced access of rice farmers to local urban markets and depressed the prices they received for their produce. Liberalization thereby reduced incomes, deepened poverty and increased malnutrition and food insecurity among rice producers. Although people generally have not been reported dying from hunger, testimonies clearly indicate that many community members do not have permanent access to adequate food in terms of quantity and quality anymore, as the realization of the right to food would require. Children and women are found to be the most affected by malnutrition. The purchase of food increasingly requires sacrifices, which limit the realization of other human rights such as the right to health and education. Vulnerability to external shocks has increased in all the communities visited for this study.

The negative impact of liberalization hits a social group, which in many cases is already marginalized because of scarce access to land, weak bargaining power towards middlemen, and weak infrastructure. Natural disasters, such as Hurricane Mitch, Tropical Storm Michelle, and droughts were other important factors, which negatively influenced the ability of rice-farming communities to feed themselves in Honduras and Indonesia. It is important to note however, that farmer's market access and incomes had already been reduced through increased imports before. As a result, natural disasters hit them harder than they should have done. Furthermore, dumped imports and excessive food aid often depressed the market longer than necessary and became an obstacle for farmers to recover from the shocks. Vulnerability to new external shocks is higher now than ever before, largely due to liberalization policies.

The case studies show that the opening of the market was a key factor for increased imports or even import surges. Trade liberalization took place in Ghana and Honduras mainly in the beginning of the 1990s and in Indonesia in 1997, and in all countries was followed by significant increases of imports. In all cases, these measures were part of structural adjustments required by the IMF and the World Bank. In these countries, the commitments of governments made under the AoA of the WTO did not affect tariffs, as they had been bound above the applied levels.

It is worth noting that governments of all three countries reacted to increased imports after 2000 by at least moderately re-regulating imports. These initiatives, which were very necessary (although not sufficient) in order to protect the right to adequate food of rice farmers, were heavily obstructed by external actors. The most striking case in this respect is Ghana, where the IMF successfully pressured the government to suspend a tariff increase only 4 days after the start of implementation. In Honduras it is DR-CAFTA, which will progressively reduce rice tariffs to zero until 2024. And in Indonesia, the World Bank is currently pushing hard for import liberalization. These facts confirm the high relevance of external pressure and the necessity to strengthen extraterritorial obligations under the right to food.

Other parts of SAPs, such as the privatization of agricultural services and credits, the liberalization of input markets and the removal of guaranteed prices and public procurement, represented a heavy burden for rice farmers in all three case studies. Farmer's access to seeds, fertilizers, machinery, extension services, and marketing facilities was substantially reduced through these measures, which meant a dramatic increase of production costs. Combined with market displacement and farm gate price depression through imports, the cuts in producer support caused drastic declines of incomes and were found to be a major reason for malnutrition and food insecurity. It is an ironical paradox that, while all these countries reduced support to domestic production, they all faced cheap imports, which in many cases were only possible because of heavy subsidization. In Honduras and Ghana for instance, the US dumping through commercial exports and misused food aid was a significant factor for import surges.

In all countries, the external and internal pressure on the governments to reduce or not to increase tariffs on rice is very high. The main argument brought forward is the interest of (poor) consumers in low prices. Case studies, however, do not confirm the expectation that consumer prices would decrease as a result of liberalization. In Indonesia, consumer prices for rice actually increased in the period of the liberalized market. In Honduras, decreasing import prices and producer prices are not reflected in low consumer prices either. The main reason in both cases seems to be the oligopolistic structure of the market, largely neglected by proponents of liberalization.

8.7 Perspectives in Times of Food Crisis

While consumers had not benefited from liberalization in the period empirically covered in the studies, recent developments make this even more obvious. The soaring commodity prices for rice in the international markets are reflected most immediately in those countries that have opened their markets most for imports. While the price increases in Indonesia, which is eager to preserve self-sufficiency, remained modest, in Honduras local rice prices in Honduras climbed by 53% between August 2007 and August 2008 (FAO 2009a, 31). The number of rice producers had declined from 25,000 at the end of the 1980s to 1,300 today. These are, needless to say, not at all in a position to increase their production quickly enough in the short term to close the supply gap that has resulted from the lack of affordable rice imports in recent months. In the case of Ghana, according to a World Bank report, prices for rice and maize have increased by 20-30% between the end 2007 and spring 2008 (Wodon et al. 2008). Ghanaian rice producers, according to the report, seem to benefit from this increase to some extent. However, after two decades of structural adjustment in this sector, they currently represent only 3.9% of the population and only cover 20% of the national consumption. The consequence is that domestic prices followed the international ones and food insecurity is sharpened, especially among poor urban consumers.

Against this background, the lobby work for comprehensive liberalization announced by the high-level task force on the global food crisis is not acceptable. In particular, following particular formulae in trade policy must not, under any circumstances, become the condition for gaining allowances or loans to combat hunger. The fact that similar practices have still been carried out by the IMF and the World Bank, even in the recent past, certainly gives cause for concern in this respect. In fact, according to the FAO, 43 states in Asia, Africa, Latin America, and the Caribbean further decreased their tariffs or custom fees on import tariffs in 2008 as a reaction to the price increases (FAO 2009a, 7). According to De Schutter, this is a problematic response because it can lead to serious losses in government revenues and, in the medium term, can encourage a further increase in imports at the cost of local producers. FAO and the World Food Programme (WFP) share this preoccupation: "High food prices have prompted the removal of import restrictions. Tariffs on food imports were reduced or eliminated in many low-income, food-deficit countries (LIFDCs). When such measures are maintained for long periods, there is a risk that they reinforce the import surges that started in the mid-1990s, with negative consequences on long-term domestic food production" (FAO and WFP 2009, 58). In fact this risk is getting higher as agricultural commodity prices have come down considerably since mid-2008 (FAO 2009b).

The study indicates a high urgency to explore and implement policy options consistent with human rights obligations of both developing and developed countries. The negative results of trade liberalization, structural adjustment, and merely market-driven agricultural policies in the rice sector and the current food crisis reveal a need for public policies and development assistance to re-establish meaningful support for local rice producers, especially smallholders. They also reveal the need for more policy spaces to re-establish public grain stocks and to protect the markets from cheap imports. Although such cheap imports have recently decreased or even stopped in many countries, they might occur again whenever international prices fall, which is very likely, given the high volatility of international markets.

This means that under no circumstances, should developing countries be obliged, through bilateral free trade agreements or the WTO Agreement on Agriculture, to reduce the ceilings for their import tariffs. On the one hand, more public support and market protection for the rice sector would help to protect and fulfill the right to food of vulnerable small-scale rice producers. On the other hand, in the long run, it would help reduce the reliance of developing countries on highly volatile international markets and thereby secure more stable prices for poor consumers.

References

Action Aid International (2005) The impacts of agro-import surges in developing countries: A case study from Ghana. Action Aid International, Accr

Ayine D (2006) A study on dumping of rice in Ghana and of possible material injury caused or threatened to the domestic industry. Commissioned by Oxfam GB, Accra (unpublished)

Assuming-Brempong S, Bonsu Osei Asare Y, Anim-Somuah H (2006) Import surge and their effect on developing countries. Ghana case study. Rice, poultry meat and tomato paste. In: Morgan N, Cluff M, Rottger A (eds) (unpublished)

De Schutter O (2008) Background note: Analysis of the world food crisis by the U.N. Special Rapporteur on the right to food. Special Rapporteur on the Right to Food, Geneva

FAO (2004) State of food insecurity in the world. FAO, Rome

FAO (2006a) FAO briefs on import surges – Countries, no. 5 Ghana: Rice, poultry and tomato paste. FAO, Rome

FAO (2006b) Statistical yearbook. FAO, Rome

FAO (2007a) Briefs on import surges: Commodities no. 2: Import surges in developing countries: The case of rice. FAO, Rome

FAO (2007b) FAO briefs on import surges no. 11. The extent and impact of import surges in Honduras: The case of rice. FAO, Rome

FAO (2009a) Country responses to the food security crisis: Nature and preliminary implications of policies pursued. FAO, Rome

FAO (2009b) Food outlook. Global market analysis. FAO, Rome

Garbers F, Gauster S (2004) La economía campesina en el contexto de la apertura comercial en Guatemala: Una aproximación después de la firma del TLC-CAUSA. CONGCOOP, Guatemala.

IMF (2003a) Country Report no. 03/133, May 2003, in: http://www.imf.org/external/pubs/ft/scr/2003/cr03133.pdf

IMF (2003b) IMF approves US 258 million PRGF Arrangement for Ghana. Press Release No. 03/66, May 12, 2003

International Food Policy Research Institute (IFPRI) (2008) High food prices: The what, who and the flow of proposed policy actions. IFPRI, Washington, DC

Japan International Cooperation Agency (JICA) and Ministry of Food and Agriculture (MOFA) (2006) The study on the promotion of domestic rice in the republic of Ghana, Progress Report (1). Nippon Koei Co., Ltd. (unpublished).

Khor M, Hormeku T (2006) The impact of globalisation and liberalisation on agriculture and small farmers in developing countries: The experience of Ghana. Third World Network (TWN), Accra

Mulyo S (2004) Indonesian rice policy in view of trade liberalization. In FAO, 2004: FAO Rice Conference. FAO, Rome

MOFA (2006) Agriculture in Ghana. Facts and figures (2005). MOFA Statistics, Research and Information Directorate (SRID), Accra

Nizwar S, et al. (2007) Policy operationalization of unhusked paddy floor price and ceiling price. Policy analysis paper. Centre for Agricultural Social Economic and Policy Studies, Bogor

Oxfam (2004) El arroz se quemó en el DR-CAFTA. Cómo el Tratado amenaza los medios de vida de los campesinos centroamericanos. Oxfam Briefing Paper 68

Oxfam (2005) Kicking down the door. Briefing Paper 72, London

Paasch A (ed), Garbers F, Hirsch T (2007) Trade policies and hunger. The impact of trade liberalisation on the right to food of rice farming communities in Ghana, Honduras and Indonesia. Ecumenical Advocacy Alliance, Geneva

Ponce Sauceda M (2003) Las Negociaciones del Tratado de Libre Comercio Estados Unidos – Centroamérica (CAFTA) y su Impacto en la Economía Rural en Honduras. Mimeo, Tegucigalpa

Ponce Sauceda M (2004) Análisis del Mercado de Arroz en Honduras y el Impacto del Dumping Estadounidense. Mimeo, Tegucigalpa

UN (1976) International Covenant on Economic, Social and Cultural Rights, adopted and opened for signature, ratification and accession by UN General Assembly resolution 2200A (XXI) of 11 December 1966, entry into force 3 January 1976, in accordance with article 27, see http://www.unhchr.ch/html/menu3/b/a_cescr.htm . Accessed 28 June 2007

UN Committee on Economic, Social and Cultural Rights (CESCR) (1999) Substantive issues arising in the implementation of the international covenant on economic, social and cultural Rights, General Comment 12, The Right to Adequate Food (Art. 11), Twentieth Session, 1999, see http://www.unhchr.ch/tbs/doc.nsf/0/3d02758c707031d58025677f003b73b9?Opendocument. Accessed 28 June 2007

UNDP (2005) Integrated assessment of the impact of trade liberalization. A country study on the Indonesian rice sector. UNDP, Jakarta

UNDP (2006) Informe de Desarrollo Humano Honduras 2006. UNDP, Tegucigalpa

Windfuhr M (ed) (2005) Beyond the nation state. Human rights in times of globalization. Global Publications Foundation, Uppsala

Wodon Q, Tsimpo C, Coulombe H (2008) Assessing the potential impact on poverty of rising cereal prices: The case of Ghana. Policy Research Working Paper 4740. The World Bank Human Development Network, Washington, DC.

World Bank (1994) Honduras: Country economic memorandum/Poverty assessment. Report No. 13317-Ho. Washington, DC

World Bank (2005) Global agricultural trade and developing countries. World Bank, Washington, DC

Chapter 9
Hunger, Poverty, and Climate Change: Institutional Approaches, a New Business Alliance, and Civil Courage to Live Up to Ethical Standards

Franz-Theo Gottwald

Abstract This article describes the problems of hunger and poverty, and how they will become ever more intractable through anthropogenic climate change. It also discusses the effects of anthropogenic climate change on agriculture and water resources. It goes on to delineate conventional and alternative approaches to fighting the problem of world hunger and to evaluate their efficacy. Then the author presents the case of Grameen Danone – a project, which is successful without development aid – but through cooperation between industry and civil society and which operates fairly and sustainably based on the ethical principles of subsidiarity and solidarity.

9.1 Hunger, Poverty, and Climate Change

According to UN predictions the next decades will see continued rapid population growth. The projected figures are seven billion by 2015 and nine billion by 2050. The twentieth century, despite triple population growth from 2.2 billion in 1950 to more than 6.5 billion today, managed to increase per capita production of food. This was mainly due to increased productivity per acre. But owing to great regional variations and unequal efforts by different countries, there has been very little change in the numbers of people starving since 1970.

Reducing hunger despite steadily growing population figures requires increased productivity as well as improved methods of distribution. This can only be achieved through effectively fighting poverty. The world community has not gotten anywhere near the goals set by the World Food Summit in 1996 which stipulated that by 2015 the number of starving people will have to be halved from 800 million to 400 million, as today's estimates put the number of people worldwide who go hungry every day at around 924 million. The Millennium Development Goals (MDG) set

---

F.-T. Gottwald (✉)
CEO Schweisfurth Foundation, Munich, Germany
e-mail: cthomas@schweisfurth.de

in 2000 indirectly lowered the target figure: The target is no longer set at half of the total world population, but at half of their share in total world population figures. This means that with an increased rate of population growth, more people may be hungry without endangering the Millennium Development Goals.

The Global Monitoring Report 2008 published recently expressed yet another dire warning that the Millennium Development Goals 2015 will not be reached due to the financial crisis of 2008 and that between 200,000 and 400,000 children may be in mortal danger. Food is becoming even scarcer in the world's poorest regions. According to researchers, incomes of 390 million of the poorest Africans will be reduced by 20%. This far exceeds the most negative predictions for the western world. The main reasons are declining prices for raw materials as well as massive reductions in investments in developing and emerging countries (http://siteresources.worldbank.org/INTGLOMONREP2008/Resources/4737994-1207342962709/8944_Web_PDF.pdf).

Hunger and poverty are predominantly concentrated in rural regions of the globe. There is a very real danger that small subsistence farmers who sell their produce locally will be squeezed out of the market, while rising prices for basic food items will increase hunger among the urban poor and indigent populations. Climate change will exacerbate exigencies of hunger and poverty.

Politically, economically, and morally, man-made global climate change has been recognized as the central challenge of the twenty-first century. The latest report issued by the UN's Intergovernmental Panel on Climate Change (IPCC) removed the last remaining doubts still entertained by science, concerning the fact that man-made greenhouse gas emissions actually have been the cause of global warming. The report gives clear indications that the global average temperature increases and its expected ecological consequences might happen a lot sooner and will be much more serious than suspected until a few years ago. It will not be our grandchildren and great-grandchildren as thought until recently, but our children who will be faced with a totally different world. Whether these changes will remain within a manageable frame, will be determined by our present-day climate protection policies and implementation.

This quickened pace of change has been due to various processes happening a great deal faster than expected. The Greenland and western Antarctica ice shelves for example are thawing a lot faster than predicted in all existing climate models.

Agriculture being especially close to nature will keenly feel the impact of climate changes. Frequent extreme weather events like droughts or storms will lead to erratic harvests and crop failures in all industrialized states. The extensive melting of alpine glaciers and possibly the much larger glaciers of the Himalayan region expected for this century, would catastrophically impact normally dependable water supplies and consequently annual harvest yields. Hundreds of millions of people in China, India, Pakistan, Nepal, and Bhutan would have to fear for their water supplies. A reduction in glacier size would also contribute more to the warming of the earth, as large light-reflecting masses of ice would have disappeared. The increasing unpredictability of the Indian sub-continent's monsoon rains is an equally ominous sign. Monsoon timing, length, precipitation amounts and distribution decide agricultural yields as well as industrial production levels and social well-being in India and other South Asian and Southeast Asian countries. Recently two man-made developments have started to impact monsoon rains: Huge clouds made up of exhaust fumes from power plants and traffic reflect the sun's rays back into outer space. This leads to a reduction of temperature differences between land and sea – and thus impacts monsoon circulation – and in a worst-case scenario may result in a total cessation. Conversely, heightened evaporation due to increased temperatures will lead to a tendency toward heavier monsoon downpours. This results in increased unpredictability and regional differences: While some regions are totally rain-starved, others experience ever more destructive inundations. This move toward less predictable monsoon rains has been observed for the past 30 years. How agriculture in the Indian sub-continent will be able to adjust to these changes cannot be predicted with any certainty.

Extensive changes in precipitation are also expected for Latin America and Africa. There will be a drastic reduction in precipitation in the Amazon region. It is foreseeable that due to clear cutting of tropical rain forests for soy production and pasture lands, 40% of the Amazonian region will have been converted to dry steppes by 2050. This would release almost as much carbon dioxide as the burning of coal, oil, and gas in the twentieth century. The warming trend of the Indian Ocean threatens an ever-increasing lack of water in southern Africa, thus worsening the periodically occurring drought crises of the past decades. The experts are still undecided if the Sahel region will suffer a similar fate – or if increased precipitation might be the result. These changes will have far-reaching consequences for food security – and they will happen a lot faster than predicted a few years ago. While up till quite recently, permanent changes were expected to take place within several decades, some have become reality much sooner. The Australian Meteorological Association for instance, has begun to classify the persistent drought conditions in some of the most important producing regions as the probable future "normal case" scenario.

9.2 Institutional Approaches Toward Sustainable Solutions

At the international level the problems and challenges listed above have been most widely known, extensively analyzed and solutions sought. A number of intergovernmental and non-governmental organizations (NGOs) have worked out recommendations and standards for a wide spectrum of topics, and developed political goals and parameters for their implementation. All these institutional approaches are value-oriented and represent a politically normative implementation of ethical discourses by target groups and stakeholders.

The UN's Food and Agriculture Organization (FAO) is the most prominent among government-supported organizations. Its main task is the creation of background knowledge on available institutions for combatting hunger. Founded in the middle of the 1960s, in an attempt to combat the problem of environmental and human pesticide poisoning, it has been developing and propagating the program of Integrated Pest Management. This institutionalized program's aim is not the total abolishment of chemical herbicides, but rather the attempt to have their use limited to appropriate levels. In conjunction with organic methods like encouraging natural predators that prey upon pests, chemicals should only be used when pest numbers exceed certain limits – and not according to a fixed schedule, regardless of need. Plant diseases are to be reduced through versatile and specific crop changes. Integrated plant protection requires more highly informed and educated farmers as compared to the system of simply applying pesticides in accordance with the calendar, which might be an impediment to its introduction in developing countries. Forty years after its introduction, the FAO in its revised Code of Conduct for the application of chemicals, charges "stakeholders" with the moral obligation of propagating the system of extensive application of integrated plant protection in developing countries. The underlying problem is, however, that this code of conduct, like practically all international instruments, is not legally binding

A more recent development is the FAO's propagated system of "Conservation Agriculture" which culminates in zero tillage. Permanent cover crops or mulching with plant materials improves humus content and reduces soil erosion. Higher amounts of humus contained in the soil mean better retention of carbon dioxide, thus helping agriculture become a carbon sink once more, especially as zero tillage eliminates high fuel costs for tractor use. Simultaneously, many farms engaged in conservation agriculture have reached higher yields. There is the danger, however, that farmers use more pesticides to control other undesirable plants growing in those unploughed fields. Therefore, extensive training in the application of appropriate cover crops and tailor-made measures of integrated conservation agriculture are required. While in the southern regions of Latin America (Brazil, Argentina, and Paraguay) around two thirds of all croplands are cultivated in this way, the rate holds at 20% in the US. Elsewhere in the world "conservation agriculture" is practically unknown. This would suggest that farmers working larger acreage have more opportunity to test the new procedure, have the means to invest in new machinery and are better able to deal with lower yields, which often occur during the changeover years. The FAO's efforts are necessarily limited to dissemination of information on standards and pilot projects in order to work toward promotion of sustainable agriculture. However, without available funds, they have to rely on outside donors and their support. Detailed and binding international standards for sustainable production systems like integrated plant protection and "conservation agriculture" are hard to define in an international context, as methods have to be adjusted to regional needs and require a national legal frame of support.

While international governmental organizations like the FAO have not set binding standards for sustainable production methods, the food security as well as animal and plant health sectors have precise and legally binding international standards. They have been defined by three specialized organizations:

- The World Health Organization and FAO's joint Codex Alimentarius Commission (CAC). It sets internationally agreed upon standards for food

additives, pesticide residues, and hygiene regulations, which assure that there are no appreciable health-risks involved. At this writing there are 165 member states.
- The World Organization for Epizootics (OIE) has been created to prevent epidemics and also to work toward improving animal health in general. They set down standards for dealing with zoonotic epidemics like hoof and mouth disease, BSE and cattle plague, with a special emphasis on setting criteria for freedom from disease and length of quarantine measures. International trade stipulates that exporting countries have to abide by the importing countries' standards – which, however, must not exceed going practices for necessary protection. The importing country has permission to check the quality of veterinary services in the exporting country. Recently, the OIE's mandate was extended to cover animal protection aspects. The OIE has 168 member states at this writing.
- The International Plant Protection Convention (IPPC) works toward the global prevention of the spread of plant pathogens. Importing countries have the basic legal right to check imports, demand disinfection measures, or prohibit the import outright. These measures must not exceed necessary requirements for risk prevention. The IPPC sets standards for certification, risk assessment, and documentation. There are 165 member states.

These standards closely focus on products posing no health risks, but usually do not take production process sustainability into account. Thus, they are only marginally useful as standards for dealing with challenges and finding solutions for global agriculture. Recently, all three organizations have displayed tendencies toward checking production processes as to their sustainability. CAC in conjunction with IFOAM has been developing standards for organic agriculture, the OIE mandate has been extended to include animal protection and an official assessment of the IPPC suggested including questions of biodiversity in its standards in the future.

The organizations are working toward internationally harmonized standards in order to remove unnecessary trade barriers for international trade. This also includes testing and evaluation standards. The standards thus defined are purely recommendatory in nature. Governments of member states are free to set higher or lower national standards, and demand compliance for imports. The founding of the WTO, however, has changed all this. The WTO's SPS agreement on Sanitary and Phytosanitary Measures is based on CAC, OIE, and IPPC in order to identify "necessary" standards in this sector.

The WTO Agreement on the Application of Sanitary and Phytosanitary Measures is designed to prevent relevant standards being misused for setting up technical trade barriers, while simultaneously allowing WTO members to maintain the desired level of protection. The SPS agreement directly corresponds to existing CAC, OIE, and IPPC standards. Trade measures based upon those standards are considered licit and though voluntary, have thus been widely institutionalized. Countries, whose producers fail to keep within these standards, have little chance of establishing export markets. Conversely, it has become more difficult for importing countries to try to set standards beyond those set by the individual organizations. In such cases, importers are saddled with the burden of proof why higher standards and thus higher levels of protection are desired.

The SPS agreement's main aim is protection from "imported" perils, and only pertains to production procedures to the extent that they directly impact finished product quality. Hygiene specifications at abattoirs would be a case in point.

But even if future CAC, OIE, and IPPC codes were applied to aspects of sustainability in production processes, it would hardly have any direct effect on the application of the SPS agreement. Trade sanctions based on production standards will remain prohibited. Experts do not expect any changes in the near future. Therefore, these organizations' standards will become the upper limit for all imported products. This highlights two problem areas of CAC work and of other organizations.

9.2.1 Voting Rules

In principle the CAC aim at consensus decision-making; if this proves unfeasible, a simple majority wins, with a one-member, one-vote procedure. Recommendations are internationally valid without restrictions even if they were passed with only a small majority vote. Voting on hormone residues in beef proves the problems inherent in this procedure: 69 countries participated in the voting, which is less than half of CAC members. 33 voted for and 29 against the proposed maximum permissible value, while ten members abstained. Nonetheless, the WTO arbitration panel used this standard to decree that the EU's import ban on beef containing hormone residues is at variance with the SPS agreement.

9.2.2 Participation of Member States in Voting Sessions and the Composition of Delegations

It has been an ongoing bone of contention for developing countries that they have been given very limited access to CAC activities (and other organizations listed in the agreement), as far as their participation in expert panel meetings and their ability to influence decisions are concerned. Industrialized countries have far larger delegations than developing countries. As a rule, the EU and the US make up almost half of all participants in the commission and expert panels. Another problem area is the influence exerted by transnational food corporations upon CAC negotiations. They are members of governmental delegations from industrialized countries and are granted observer status during meetings in their capacity as representatives of international industry associations. During the 19th meeting session, one quarter of participants were directly linked to industry. Only 1% were NGO representatives serving public interests like consumer or health protection. The highly delicate task of the Codex Alimentarius of setting balanced international standards serving both health protection as well as easing trade regulations, is not made any easier due to the overpowering influence of lobbyists working mainly for free trade.

Another international, value-driven approach toward sustainable solutions is ecological agriculture. Ecological agriculture has the longest tradition of defining and supervising extensive agricultural production standards. Its international governing body IFOAM (International Federation of Organic Agriculture Movements) defines basic principles and general benchmarks for worldwide organic agriculture:

- The Principle of Health: Organic Agriculture should sustain and enhance the health of soil, plant, animal, human, and planet as one and indivisible.
- The Principle of Ecology: Organic Agriculture should be based on living ecological systems and cycles, work with them, emulate them, and help sustain them.
- The Principle of Fairness: Organic Agriculture should build on relationships that ensure fairness with regard to the common environment and life opportunities.
- The Principle of Care: Organic Agriculture should be managed in a precautionary and responsible manner to protect the health and well-being of current and future generations and the environment (http://www.ifoam.org/about_ifoam/principles/index.html).

International guidelines have been adapted and refined by regional and national member organizations in order to create specifications tailored to local requirements for organic agriculture. Organic agriculture not only aspires to producing pure food and greater biodiversity, but in a similar vein to "conservation agriculture," higher shares of humus in the soil and more carbon sequestering. The FAO in its conference report on organic agriculture and food safety held in conjunction with IFOAM, states that organic agriculture has a number of advantages:

The entire eco-agrarian system is more solid due to higher humus content of soil. This entails better water retention and reduces the need for irrigation, while increasing yields during periods of drought. In comparison to conventional agriculture its climate footprint is positive. Many studies have shown that fossil fuel use for fertilizers is one third lower; soils sequesters twice as much CO_2, while nitrogen or nitrous oxide emissions are also reduced. In contrast, methane emissions due to animal husbandry and rice cultivation are approximately equal. "Conservation agriculture" with its zero tillage can show better greenhouse gas balance per surface unit or food unit. These emissions, however, do not take into account the normally required amounts of pesticides and artificial fertilizer.

Organic agriculture practiced in low-yield areas can lead to immediate increases in productivity, especially if the areas had not been cultivated in a sustainable fashion. In these cases, the World Bank recommends either organic agriculture practices or methods that closely resemble those of organic agriculture. The biggest obstacles for the application of these methods in developing countries is the fact that they are "knowledge-based" and require a relatively high amount of consultation and a minimum of education. In favored areas in Europe and the US, somewhat lower yields and higher production costs are to be expected in comparison with conventional cultivation. Therefore, in order to be profitable for farmers, organic products will have to be sold at a premium.

This necessitated reliable labeling of organic food products right from the start. At the beginning, independent associations created and administered appropriate food labels and methods of certification. Although the surveillance and control mechanisms worked well, the multiplicity of varying labels depicting very similar, if not totally identical standards, were soon seen as a major obstacle to increasing market share of organic products. As terms like "organic" and "ecological" had no trademark protection, "copycats" managed to mark conventional food items with the same labels, which confused consumers and did not help to build consumer confidence. As of 1991, terms like "organic" and "ecological" have been protected by the EU and may only be used for food items produced and certified in compliance with appropriate EU regulations. These standards conform to European organic agriculture associations, and thus are commensurate with IFOAM standards, with a few exceptions where these strict standards have been eased a bit, for example by allowing operations to certify individual units – like crop cultivation – while others like animal husbandry conducted in the same location, would still be conventional. As the EU and the US are the biggest markets for organic agriculture, their defined standards carry the status of global guidelines. The EU is at the cutting edge, especially as the EU-defined standards are compatible to those of IFOAM.

An international fair trade ethos has been created to serve as an additional semi-institutional structure in the globalization of food production and distribution. Around 800 world stores run mainly by volunteers and 8,000 activist groups carry on the fair trade movement in Germany alone, which started out in the early 1970s to support the cry for a "New World Trading Order." Trade was limited to produce like coffee, tea, cocoa, or bananas on the one hand and arts and crafts on the other. Independent small producers, preferably united in democratically structured cooperatives have been the trade partners of choice. On the surface, there are no conflicts of interest between owners and employees. Problems like hiring seasonal workers for small-scale farming are rarely addressed by fair trade. Social goals are met through higher prices paid to producers (or organizations).

Additionally, further criteria for business dealings between growers and traders have been established, for example long-term trade relations, (low interest) advance financing and help in product development. Once fair trade ideas had been extended to products from privately owned large plantations growing tea or bananas, social standards gained in importance. As a rule standards for basic labor rights defined by the International Labor Organization (ILO) are applied. Higher prices levied for social purposes have remained an integral part.

In June of 1998 the "Declaration on basic principles and rights at work" was passed. It identified most of the 182 ILO conventions and called on member states to ratify them. The "basic labor standards" or "basic rights at work" read as follows (BMZ 2000):

- Right of association and right for collective bargaining (Conventions No. 87 and 98),
- Equal pay and nondiscrimination at work (Conventions No. 100 and 111),
- Prohibition of forced labor (Conventions No. 29 and 105),
- Prohibition of child labor (Conventions No. 138 and 182).

The umbrella organization of fair trade stores revised their "Criteria for Fair Trade" in 1999 and included a list of demands addressed to all participants in the trade chain. In addition to trade, one important criterion is active participation in information and educational work pertaining to producers and questions of trade policy. The TransFair seal guarantees that producers maintain fair trade criteria. They have to prove their socially exemplary conduct to certification authorities, while retailers may apply the seal to fairly traded goods only. There are no further requirements imposed on retail outfits on how to conduct their business beyond paying a higher price, which allows producers to supply the TransFair mandated social benefits and a licensing fee for the seal. These injunctions do not apply to general purchasing policies, which allow discounters like Lidl to offer products bearing the TransFair seal.

9.3 Grameen-Danone – A New Business Solution

Besides the myriad political, institutional, and organizational approaches toward warding off the horrors of hunger, poverty, and climate change, business and industries are establishing business solutions within civil societies and without recourse to politically organized legal frameworks of help or development aid. These solutions are imbued with values like the precaution principle and responsibility or the ethos of sustainable development. This approach is best represented by the Grameen–Danone Alliance in Bangladesh.

A joint venture was formed between Danone and four branches of Grameen, i.e. Grameen Byabose Bikash (a management counseling company), Grameen Kalyan (social services), Grameen Shakti (energy), and Grameen Telecom (a nonprofit enterprise holding a major share in Grameen Phone which in turn is the largest mobile phone company in Bangladesh).

These businesses are run as social enterprises, i.e. as defined by Muhammed Yunus these are companies that show neither losses nor distribute dividends. Grameen Danone Foods Ltd. gives a symbolic dividend of 1% which is to be reinvested in order to integrate local communities into production and distribution projects. Local entrepreneurs receiving help in improving their production methods are the main beneficiaries. They are supported in setting up cost-effective and labor-intensive production models, which in turn help create more jobs in the distribution sector.

Grameen Danone Food Ltd. coordinates four main groups of stakeholders: customers, competitors, shareholders, and employees. Customers receive yoghurts that are tailor-made for their needs at a good price. As the other Bangladeshi dairies are Grameen Danone Food Ltd.'s main competitors, GDFL make sure that small farmers who often own no more than one cow are part of company planning and encourage them to expand their operations. Danone's shareholders profit from the "lofty goals and ethically praiseworthy image" being established with GDFL's customers in wealthier countries. They also run no risk, but nonetheless are assured their 1% dividend, which translates into a no-loss, no-risk venture. Employees are assured of a secure place of work in keeping with ILO social standards (no child labor, no forced labor, no discriminatory practices, freedom of assembly, maintenance of health standards, adequate remuneration, decent working hours, etc.). The company started out with 20 workers and has grown to around 50. There are also 40 "Grameen Ladies" who take care of deliveries to surrounding villages.

Grameen Danone Food Ltd. processes milk from surrounding farmers. Sugar and molasses are likewise from local growers. And also the ingredients for packaging are made from organic and biodegradable corn starch.

Supplier and employee security is also well regulated. Workers are under contract by Grameen Danone Foods Ltd. Farmers receive suggestions on increasing and improving their output. They, too, enter into fixed contracts with the company. The "Grameen Ladies" buy their yoghurts for five Takas and sell them for six, thus giving them a small profit margin per sale. Before they start their rounds, the women attend two-day-workshops to learn about refrigeration, ingredients, and health aspects (Yunus 2008).

9.4 Subsidiarity and Solidarity as a Social Ethics Principle

The example of Grameen-Danone Food Ltd. is based upon a deep understanding of two social ethics principles which are essential for sustainable development: Subsidiarity and solidarity.

In times of liberalization, the principle of subsidiarity is particularly useful for basic orientation and assessment of the suitability of international agreements and civil society approaches toward implementation of the right to food. It judges individual or group liberty and responsibility according to ethical standards and gives preferential treatment for example in the search for solutions in the fight against hunger to individual, local or regional community actions in contrast to states acting on national or international levels.

The ongoing globalization of the food sector, inclusively the new and increasing power and concentration among enterprises and corporations, is not only the result of deregulation and more "free trade," but also the consequence of subsidiarity not having been raised to a political formal principle of globally active organizations and institutions. Otherwise, ways and means of giving preferential treatment to lower echelons of nation states as well as countries, towns, and communities would be available where economic activists are active or where they wield considerable influence. Subsidiarity encourages higher institutions to withdraw and allow "subordinate groups" the freedom of implementing food equality in keeping with local and regional self-determination.

Many studies have shown that there would be enough local and regional power for food self-sufficiency and the service of local and regional markets, but due to misinterpretations of globalization and misdirected trade this cannot come to the fore (Ali-Dinar 2009, Dorward 2004, Jawara and Kwa 2004). Rarely – and usually due to climate change, inclement weather or other natural disasters like major fires – have subordinate groups, i.e. villages, towns, and regions been unable to master challenges relating to feeding their inhabitants on their own. This leads to the stipulation that an eco-social structuring of the future of alimentation must take the principle of subsidiarity seriously, making sure that state as well as civil society activists or higher organizations limit their activities to support and refrain from interfering with supply, distribution, or trade challenges.

The farm surpluses of the rich countries of the industrialized North are still used to pry open markets or destroy existing ones in the developing world – which is in stark contrast to the idea of solidarity. Organizations in countries should receive subsidiary assistance, which furthers their independent development, thus helping local agriculture and food production become self-sufficient. It is a proven fact that stabilizing and expanding local production and processing economies will markedly improve supply as well as effectively fight poverty and hunger. This would definitely be the path of choice. Agrarian exports from industrialized countries or from emerging countries into developing countries impede these countries' development just as much as rampant exports of raw materials for animal feed and fuel plants from these countries. (Fischler 2007).

If the principle of subsidiarity were used as a guideline, it would follow that the WTO should be given the power of granting and guaranteeing exceptions for basic food items. This tool could be introduced in developing countries and slowly, but steadily help apply their local ecological, social, or cultural advantages. Assuming responsibility on national, regional, and local levels would act as an incentive for exploring a multitude of new development ideas and would also help against the disproportionate industrialization and centralization in food economies.

If the subsidiarity principle were to guide political frameworks, it would facilitate and encourage the enforcement of rules for eco-social "regional market economies." On the basis of minimum regulation frameworks against competitive distortion, the subsidiary option for regionalizing agriculture and food production is in step with world consumers' as well as small and medium size producers' growing regional consciousness (Schweisfurth, Gottwald, Dierkes 2002).

However, as noted above, there are emergency situations where in keeping with a second auxiliary guideline, solidarity has to be shown. In view of dramatic changes in climate and the number of catastrophes leaving famine in their wake, the global food security net established by the international community will have to become more active. World food emergency supplies will have to be increased considerably, while food aid will have to be scrutinized with regard to its helping as well as market destroying impacts.

The principle of solidarity can also be used to give credence to institutional measures geared toward opening up opportunities and assisting self-help measures as well as access to world trade. One way would be extending consulting services for the establishment of valid environmental, social, and animal protection standards, by putting into practice solidarity with all life forms and thus opening up new markets and trading opportunities.

The idea of solidarity, quite successfully put into practice within the EU (2007), if expanded to include the rest of the world, could lead to new social safety nets for the poor in rural areas far from civilization. Initial developments along the lines of micro-finance and micro-insurance, have shown that differentiated ways of participation can be instituted, if well-off members of society are willing to contribute. Low income rural families, living off subsistence economies, will have to be included in monetary security systems supported by the economic might of thriving operations.

An additional dimension of solidarity deals with cooperation within local or regional farming, processing, and merchandising. Good examples of cooperation between families or operations in sharing farm implements, or associations providing farmhands in times of need, show that institutionalized farm cooperatives provide more economic advantages in and for rural areas that go way beyond the idea of the commons.

9.5 Civil Disobedience – Or How to Live up to Values, Standards, and Treaties

Subsidiarity and solidarity are two key ethically grounded features designed to guide public administrators as well as actors of civil society in living up to existing conventions, standards, and treaties that have met with obstacles in compliance during day-to-day administrative life.

The FAO's "Code of Conduct for Responsible Fisheries," for example was designed to stop overfishing of world oceans. The guidelines have been ratified by 53 countries representing 96% of global fisheries. Biologists from several universities and WWF, the nature conservation organization, have proven that none of those countries fulfill all of the regulations contained in the code, while only nine have exceeded 50%. Thus, the practices of overfishing, killing of by-catches as well as environmental destruction have continued practically unimpeded (FOCUS 2009).

This is not a solitary observation. The obstacles in the path of reaching MDG have been discussed above. Likewise upholding the Convention on Biological Diversity (CBD) is far from being a reality due to a lack of legitimate national measures of enforcement. The problems of ownership and laws governing respective rights of neighbors in the case of seed and breeding stock as well as overdue revisions of laws governing property rights (e.g. an ongoing process in the PR China) and the worldwide dearth of implementation and enforcement of nature and environmental protection standards constitute an additional political arena contributing to a half-hearted fight against poverty and hunger in this age of climate change.

It becomes apparent that moral indignation and outrage in the face of hunger and poverty is similar all over the world. This is partly based on religious, partly on ideological traditions of care, (spontaneous) assistance, solidarity to name only three guiding values which make up a significant part of public consciousness. Unfortunately factual adherence to norms, standards, and conventions which would create conditions for life and commerce without hunger and poverty, has not been achieved. The year 2008 saw food riots, starving populations went on the rampage and organized hunger marches in many regions of the world (The Independent 2007).

Hence, mechanisms of enforcement have to be established to warrant freedom of choice of sustainable food practices. It is one of the tasks of agricultural and food ethics to identify reasonable ways of ensuring fair access and just distribution of food and drinking water in compliance and practical adherence with rules set up with the help of established and competent instruments (cf. Follet 2009).

Which measures would be necessary to moderate all the heterogeneous interest-driven intentions and (cultural) valuations by differing stakeholders toward a solution and toward the establishment of a culture of cooperation culminating in factual compliance with ethically based commitments?

In Europe the practice of integrating stakeholders into the process of establishing standards and setting of ethical foundations for sustainable solutions in conformity with agricultural and food requirements has mainly been tried on the national level, but also successfully in the EU. One example would be the EU project "Ethical Bio-Technology Assessment for Agriculture and Food Production. A Guide for Users" (Beekman et al. 2002). This project identifies various tools for designing frameworks of normative discussions and exemplifies the successful application of these principles, among others in societal dialogues.

It can be said that the more stakeholders have been integrated into processes of institution building designed to assure food security and fight poverty, the more closely these become internalized. Research in game theory on cooperation gains confirms these findings. These gains result from the formation of stakeholder coalitions which ended in mutual decision making in favor of sustainability (Krysiak 2004).

European Union Europe is ruled by the rule of law and the rule of the welfare state based upon cooperation and conflict mediation and stakeholder integration. This system allows people to demand living conditions guaranteeing a dignified lifestyle free from hunger and poverty as much as this is possible, even if the "Right to Food" has not been set down in European national constitutions. Conditions are quite different outside of the EU. As mentioned above, cases of insufficient and life-threatening food supply or drinking water are on the rise. All over the world, and also in Europe, regional and climatically adapted food production and processing systems cannot always be taken for granted. Two examples would be the so-called "Green Genetic Engineering" and novel biotechnological test systems that recombine genetic material from different animal species, which small farmers and consumers perceive as threatening to their existence.

A longstanding battle against green genetic engineering or fight for unpolluted seed material has been conducted by individuals like Percy and Louise Schmeiser or Vandana Shiva, and organizations like "Save our Seeds." They exemplify the need for civil disobedience in situations that are perceived as life-threatening. They also represent an example of one type of last-ditch-effort left to civil society activists if basic rights, basic values or existing standards and agreements are violated or broken on national or international levels.

The increasing domination by the agro-industrial and techno-political systems, and the structural power of displacement upon small and mid-size regional food security and food sovereignty efforts exerted by them imbues civil disobedience with new meaning. Adversaries of genetic engineering typically express their resistance or civil disobedience through field sabotage. Civil disobedience is an expression of civil society watchfulness, testing if justice is actually served (by varying types of cultivation and manufacturers of seed and breeding stock) through positive law. Whenever civil society activists recognize and are able to prove that the rule of law is not sufficiently in keeping with ethical legitimacy, civil disobedience is called upon.

A collision of positive law on the one hand and constitutionally guaranteed basic rights on the other, is possible in a modern constitutional state. This may become a challenge calling for civil disobedience. This may be the case if nations react to globalization, migration, terrorism, and home security by enacting laws which infringe upon people's dignity or basic human rights while fighting against threats. Civil disobedience in this conjunction is an expression of a collision between basic rights and positive law within existing basic laws, but not as is often cited, a conflict between the rule of law and high-minded do-gooders.

Civil disobedience aims at "realizing what is humanly possible, but has not been achieved yet" within a legal framework. Its ethical legitimacy is based upon supranational, constitutionally guaranteed basic rights, from personal conscience or from divine laws. This, however, does not constitute legal legitimization for disobedience or resistance. Civil disobedience always means defiance of existing laws and thus a breach of law. In a democratic nation ruled by law, a breach of law can never become a legally protected interest. Its positive function lies elsewhere: civil disobedience may become an "ultima ratio," a last resort, for pointing out a law's lack of conformity with basic laws and demanding redress, once all other legal means have been exhausted. Seen in this light, civil disobedience is an expression of a civil and democratic assumption of responsibility, indispensable for any constitutional state, as it positively and constructively helps perfect its legal system. This form of improved justice attained through civil disobedience, can only be achieved internally through an improvement of the legal system, not at variance with it (Garstecki 2005).

In the fight for food security and clean drinking water civil disobedience has proven a tried and true means of introducing these short and long-term interests into the power struggle within the whole of society. In the twenty-first century, civil society activists will increasingly (have to) resort to civil disobedience in their struggle against agro-industrial and agro-political misuse, if the tools for guaranteeing food security prove ineffectual and in need of empowerment from below.

References

Ali-Dinar AB (2009) Food security and food self sufficiency in Africa, *United Nations Economic Commission for Africa* (Uneca). http://www.africa.upenn.edu/ECA/FoodIntro.html. Accessed 11 March 2009

Beekman V, et al. (2002) Ethical bio-technology assessment tools for agriculture and food production. Final report bioethical bio-TA Tools (QLG6-CT-2002-02594). Agricultural Economics Research Institute (LEI), The Hague

Dorward A (2004) Agricultural liberalisation in sub Saharan Africa. Department for International Development, Imperial College http://www.research4development.info/PDF/Outputs/EC-PREP/AgriculturalLiberalisationAfricaSummary.pdf. Accessed 11 March 2009

European Communities (2007) The committee of the regions and the implementation and monitoring of the principles of subsidiarity and proportionality in the light of the Constitution for Europe

Fischler F (2007) Ökosoziale Entwicklungen gestalten. Das Beispiel der europäischen Landwirtschaftspolitik. In: Gottwald F-Th, Fischler F (eds) Ernährung sichern – weltweit – Ökosoziale Gestaltungsperspektiven. Bericht an die Global Marshall Plan Initiative, Murmann Hamburg, pp. 24–49

FOCUS (2009) Fischereikodex stoppt Plünderung der Meere nicht. FOCUS 7:58

Follet JR (2009) Choosing a food future: Differentiating among alternative food options. J Agric Environ Ethics 22:31–51

Garstecki J (2005) Ziviler Ungehorsam – Rechtsgut oder Rechtsbruch? Lecture presented at Landgericht Leipzig, 10 November 2005

Gelauff G, Grilo I, Lejour A (2008) Subsidiarity and economic reform in Europe. Springer, Berlin

Held D, Moore HL (2007) Cultural politics in a global age: Uncertainty, solidarity and innovation. OneWorld, Oxford

International Federation of Organic Agriculture Movements (IFOAM) http://www.ifoam.org/about_ifoam/principles/index.html. Accessed 11 March 2009

Jawara F, Kwa A (2004) Behind the scenes at the WTO: The real world of international trade negotiations, updated ed. The real world of trade negotiations/Lessons of Cancun. Zed Book Ltd., London

Krysiak F (2004) Sustainability with uncertain future preferences. Environ Resour Econom 33:511–531

Taylor J (2007) How the rising price of corn made Mexicans take to the streets. In: The Independent. http://www.independent.co.uk/news/world/americas/how-the-rising-price-of-corn-made-mexicans-take-to-streets-454260.html. Accessed 13 March 2009

Yunus M (2008) Creating a world without poverty. PublicAffairs, New York

Schwartz JM (2006) The future of democratic equality: Rebuilding social solidarity in a fragmented America. Routledge, New York

Schweisfurth KL, Gottwald F-Th, Dierkes, M (2002) Toward sustainable agriculture and food production. A vision of food production, processing, and marketing. IFOAM, Bonn

Watkins K, Montjourides P (2008) Global Monitoring Report 2008, United Nations. http://siteresources.worldbank.org/INTGLOMONREP2008/Resources/4737994-1207342962709/8944_Web_PDF.pdf. Accessed 10 March 2009

Chapter 10
Food Versus Fuel: Governance Potential for Water Rivalry

Lena Partzsch and Sara Hughes

Abstract Biofuel and food crop production compete for scarce arable land and water. The Mexican "tortilla crisis" in 2007 publicly revealed this dilemma. Such intersections can create challenges for developing policies – both at the global and national levels – that secure affordable and accessible food sources. Further, as water resources continue to be stretched, tradeoffs for consumptive uses will become increasingly common. Virtual water accounting is a tool that has been developed to increase our understanding of the way water is used in the production of goods, and particularly how this affects the global distribution of water through trade in these goods. The chapter presents the current understanding of global water resources and the impacts of trade on their distribution. Many linkages exist between energy security, agricultural trade, and water resource sustainability. As our understanding of the complexity of energy development and food ethics increases we will need to utilize tools such as virtual water accounting to inform policy making and to incorporate a wider array of social and environmental goals.

10.1 Introduction

Worldwide rivalry between biofuel and food production has become obvious. The United Nations Special Rapporteur on the Right to Food Jean Ziegler is demanding an international 5-year ban on producing biofuels to combat soaring food prices (Ziegler 2008). Crucial factors behind this competition are scarce arable land and water. Expanding agriculture to meet countries' growing demand for biofuels could place extreme demands on global water resources. Clean and reliable water resources are necessary for nearly all social-industrial processes, including agricultural production, the sector which is currently the world's largest user of water

---

L. Partzsch (✉)
Department of Economics, Helmholtz Center for Environmental Research, Leipzig, Germany
e-mail: lena.partzsch@ufz.de

resources (Global Water System Project 2005). Global water demand has continued to rise in nearly every sector over the last 20 years despite limited supplies, particularly in the world's expanding arid and semi-arid regions (Dregne 1986). Therefore, agricultural and energy policies that are seemingly unrelated to water use, such as institutionalized support of biofuel production, can have major water-related impacts. The discussion surrounding biofuels has been largely positive when focusing on their potential for reducing green house gas emissions, improving countries' energy security, providing economic opportunities in the world's impoverished rural areas, but – on the negative side – competing with food crops for scarce arable land. Since the Mexican "tortilla crisis" debates have become more intense (von Geibler 2007, p. 5). Because of the energetic use of import maize, the price for maize meal increased and tortillas became suddenly three times more expensive in Mexico in early 2007. Beer prices in Germany ticked upwards partially due to the increased production of biofuels, and Italian pasta has become more expensive (Spiegel Online 2007). Replacing 5.75% of Europe's fuels with biofuels by 2010, as Europe has pledged to do, or 15% of US gasoline use by 2017, as proposed by President Bush, will place enormous demands on existing cropping systems. The US plan alone would require 35 billion gallons of alternative fuel: the equivalent of 13.5 billion bushels of corn (using current technologies). Vast amounts of water are needed to produce the crops. It takes about 4,560 l of water to produce 1 l of ethanol (Jungbluth and Rohwetter 2007). The water-related consequences of large-scale biofuel production and the potential need for policy guidance in this area have yet to be fully explored.

This chapter has two major goals. The first is to establish that, in the context of the "global water crisis," a multidimensional challenge facing society, virtual water accounting is a useful tool with which to evaluate the international impacts of producing and trading biofuel stocks. Second, we use these functional links among those global institutions governing water resources, biofuel production, and trade in agricultural commodities to propose three policy solutions that would help move the international community toward a more integrated understanding of the development of biofuels within the context of the global water crisis.

10.2 What Is the Global Water Crisis?

Mounting evidence has prompted an almost universal declaration of the existence of global water scarcity, or a "global water crisis" (UNEP 2001). This is particularly true when trends are examined with an eye toward the future, as water shortages experienced by humans and ecosystems throughout the world often impose serious risks to long-term sustainability of linked socio-ecological water systems (Gleick 1998, UNEP 2006). Today, some 1.1 billion people in developing countries have inadequate access to water, and 2.6 billion lack basic sanitation (UNDP 2006, p. 2). Several studies have concluded that the world's freshwater ecosystems have already been significantly degraded in form and function due to water overuse and contamination (Postel 2006, Baron et al. 2002). The GIWA report initiated by the

United Nations Environment Programme (UNEP) predicts that freshwater shortage will increase in severity in over two-thirds of the freshwater systems by 2020 (UNEP 2006, p. 35).

At the root of the water crisis are inappropriate economic incentives for water use (Johnson et al. 2001) and insufficient social institutions and legal frameworks for water management (Salman 1999, Young 2002). Whether scholars agree with the trend or not (see Conca 2006), the global community is playing a significant role in the outcomes of local water scarcity issues. For example, international aid and development strategies have increasingly focused on addressing the critical nexus between global water scarcity and poverty, allocating funds for this problem at local levels and further integrating global economic and political networks with water scarcity issues (UNEP 2001, Rockstrom et al. 2005, Unver et al. 2003). The Millennium Development Goals challenge the world to decrease by one-half the proportion of people without access to potable water and sanitation by 2015 (WSSCC 2000). However, as Johnson et al. (2001) recognize, "[W]ater is not only becoming scarce because of increased demand, but also because of higher pollution levels and habitat degradation." Agrochemicals put particular pressure on freshwater resources by contributing toxic runoff to receiving water bodies (Bringezu et al. 2007, SIWI 2007). Fertilization and pesticide use on large plantations can lead to eutrophication of water courses, lakes, and seas, the impacts of which stretch far beyond the field.

Water scarcity is also fundamentally a product of geographic and temporal distribution, with too much water in some places at a given time and too little in other places and times. Many countries in sub-Saharan Africa or the Middle East, for example, must continually place water scarcity issues at the top of their policy agendas regardless of the existence of a current crisis; other areas around the world struggle to cope with seasonal flooding and storage issues as well as groundwater intrusion (Fig. 10.1) (Allan 1998, Turton et al. 2003).

10.3 Virtual Water: Rethinking a Resource

Understanding how water is "naturally" distributed is just the first piece of the puzzle, because global trade in agricultural and other commodities influences how water is consumed among countries. International commodity trade has an under-recognized role in redistributing global water resources. This is especially true for trade in agricultural products, which often require significant quantities of water to grow. The volume of water that farmers use to grow a crop, which is not physically embodied in the final product, is dubbed the crop's "virtual water" content (Allan 1998). Virtual water first came to the attention of scholars through Tony Allan's claims that an important reason wars have not been fought over scarce water resources in the Middle East is in large part due to the region's dependence upon outside suppliers of water-intensive agricultural products. The arid region's water requirements are not trivial. The OECD Observer reports in 2006 that, "the overall water required in the farm products imported into the Middle East and North Africa

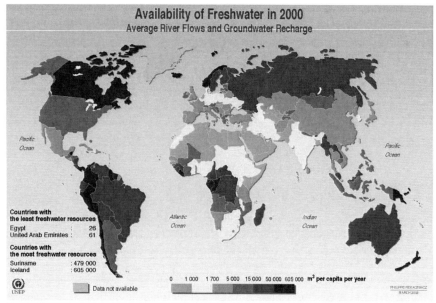

Fig. 10.1 Global distribution of per capita freshwater resources (Source: WRI 2000)

is equivalent to the annual flow of the Nile upstream of Aswan." Virtual water also provides insight into the mismanagement of water that export trends in dry regions encourage. For example, some water scarce regions of Africa are not net virtual water importers, as one might expect, but net exporters regardless of existing water scarcity in that region (Fig. 10.2) (UNESCO 2006). These trends in virtual water movement should be understood as water managers and development programs work to encourage water conservation, redistribution, and quality improvements.

Today, research has advanced to the point that virtual water accounting can be used as a useful empirical tool. For example, Hoekstra and Hung (2005) and Aldayai et al. (2008) quantified the flows of virtual water contained in thirty-three crops traded between and among the 14 regions of the world (Table 10.1).

They used adjusted ideal-crop scenarios to estimate the water required to grow each type and applied these to trade flow data within and between the regions, using generalized crop requirement data as well as site-specific information on evapotranspiration and yields. The findings are a foundational building block for quantifying the flows of water between countries as well as explaining the social, economic, and environmental consequences of the trends. They can even be used to identify policy levers: specific crop trading practices that would put undue burden – or provide relief – in instances of water scarcity. Virtual water accounting emphasizes the important role of social and political institutions – in addition to the relative availability of water resources – in determining how water is used and distributed around the world (Biswas 2004).

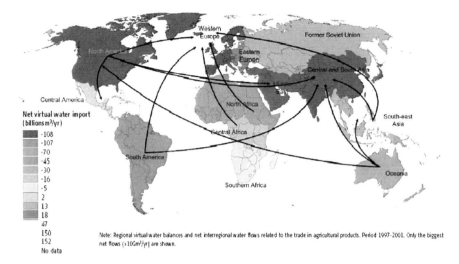

Fig. 10.2 Net virtual water imports around the world (Source: UNESCO 2006)

Table 10.1 Global virtual water flows between nations by product (adapted from Hoekstra and Hung 2005)

Product	% Global virtual water flow
Wheat	30.2
Soybean	17.1
Rice	15.4
Maize	8.9
Raw sugar	7.2

10.4 The Rise of Biofuels

Countries around the world are increasingly employing (or exploring) biofuels as cleaner, more secure alternatives to gasoline in meeting their transportation energy needs. Current biofuel technologies rely on converting crops that farmers have traditionally grown for feed purposes (i.e., corn, soybeans, sugar, and palm) into ethanol or biodiesel that could displace fossil fuels in motor vehicles, a significant source of carbon dioxide emissions (Schipper and Marie-Lilliu 1999). Ethanol is produced from crops with high sugar content such as wheat, beets, and sugar cane. These sugars are fermented into ethanol either by biologic or chemical means; biodiesel is made from oil crops such as rapeseed, soybeans, and jatropha and biodiesel from vegetable oils can be blended with oil-based diesel or used directly (IPC 2006).

Using these crops as energy stocks is being used as a strategy for meeting Kyoto Protocol commitments for greenhouse gas emission reductions, decreasing air pollution for domestic reasons, and/or generating greater domestic energy security in nonoil producing countries (IPC 2006). Another reason for encouraging growth in the biofuels sector is to revitalize a deteriorating agricultural sector, a significant concern for both developing and developed countries (Hazell and Pachauri 2006). Some farmers see biofuels as the answer to often inaccessible and unpredictable global agricultural markets, believing that, for example, "the whole bioethanol revolution will save maize farmers in South Africa (http://www.irinnews.org)."

Many of these crops are traditionally grown under rainfed conditions, but expanding production may push crops into more marginal areas that require irrigation. Relying on irrigated crops to produce biofuels would put significant strain on global water resources. The global average virtual water content of wheat, sugar cane, and rapeseed is 1,300, 175, and 1,600 m^3/ton, respectively (Chapagain and Hoekstra 2004). One liter of ethanol alone requires up to 4,560 l of water (Jungbluth and Rohwetter 2007). In addition to production, the processing of these crops can contribute major pollution loads to aquatic systems, and hence creates the potential for an environmental burden shift to producing countries when processing falls on their shoulders as well (Berndes 2002, 262).

10.5 Trade and the Distribution of Biofuels and Water

As shown in Table 10.1, four of the five crops most responsible for the global flow of virtual water are also used in the production of ethanol and biodiesel; global trade in wheat alone is responsible for 30% of global virtual water flows (Hoekstra and Hung 2005). If the trade in these commodities is to greatly increase in the future to meet the growing demand for biofuels it seems more than reasonable to propose an assessment of how this will impact the water resources in producing and consuming countries *and* to address these concerns in the policy frameworks which will undoubtedly arise around the trade regimes. New biofuel crops may add themselves to the list of "top five" virtual water crops, and those crops currently responsible for most of the world's virtual water trade will continue to grow in importance.

International trade in biofuels or in their ingredients causes additional virtual water flows, above and beyond existing crop trade and its impact on global water resources as trade volumes increase. If, for example, South Africa increases biofuel exports it will increase its net virtual water trade deficit (Fig. 10.3). The United States is likely to export less virtual water due to increased domestic demand for biofuel consumption: more of the crops traditionally exported can now be consumed domestically. Just as water is differentially distributed around the globe, so is the capacity for countries to increase their production of biofuels. The US and Brazil, currently the world's leading ethanol producers, are also major consumers, so trade in ethanol is small relative to production (Hazell and Pachauri 2006). However, countries such as China, India, Senegal, and South Africa are quickly becoming significant players in the global arena (Gonsalves 2006, IRIN News 2006). Brazil represents 50% of global ethanol exports, sending its biofuels primarily to the US and India (UNCTAD 2006) (Fig. 10.4). Trade in cane sugar and maize have not

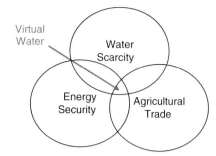

Fig. 10.3 Virtual water accounting as the nexus within global water scarcity, agricultural trade, and energy security

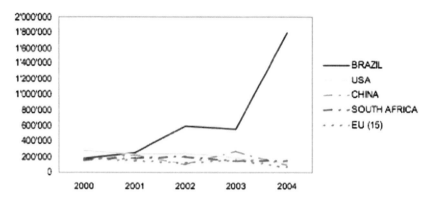

Fig. 10.4 Exports of undenatured ethanol from the five leading world exporters (in tonnes). Source: UNCTAD calculations based on COMTRADE

risen along with the ethanol boom, most likely because subsidies in developed countries favor domestic ethanol production. Biodiesel, on the other hand, represents an emerging new force in oilseeds trade; for example, palm oil exports from Malaysia and Indonesia to the EU have spiked, most likely to meet biodiesel demand.

Intensive studies conducted by the Wuppertal Institute in Germany and the Stockholm International Water Institute found that the commitments of the European Union to increase its biofuel use by 30% by the year 2025 will require imports of biofuel feedstocks from other countries (Bringezu et al. 2007, SIWI 2007). Even without considering the needs of other countries, it is not clear that there will be enough available arable land not already in production to meet the kinds of biofuel target demands set by the European Union and the United States unless significant changes in crop selection and cultivation are instituted (IWMI 2007). Implementing such changes would require political and technical exchanges (such as subsidization policy changes and technology transfers) which have thus far been avoided in the international arena. These land-use challenges do not consider the competition for water resources that biofuels may impose. Innovations such as the use of jatropha, a plant able to grow quite easily in arid regions, in biodiesel production has the potential to significantly reduce the volume of water resources needed to sustain biofuel production [see, for example, India's development of jatropha plantations within its own borders and soon in Senegal (www.jatrophaworld.org)]. However, current practices indicate that water availability will be a problem

Beyond land and water constraints, there are other concerns regarding increases in the production of biofuels. A large group of international NGOs (nongovernmental organizations) feel that the likelihood of developing countries becoming the primary producers of biofuel feedstocks due to wealthy lifestyles, while developed countries are the primary consumers, may create a scenario reminiscent of earlier colonialism based on resource extraction and exploitation (Global Forest Coalition 2006). While short-term benefits to the agricultural sector in producing countries are very likely, if the largest markets are primarily located abroad, in countries such as the United States and in the European Union, these groups worry that poorer countries will continue to be dependent on primary resource exports and may forgo opportunities to invest in increases in food production for their citizens. However, some see biofuel expansion in developing countries as a way to help reduce commodity dependency in these countries and fuel infrastructure development that will help with the distribution of staple food crops as well (UNCTAD 2006). In addition, there is concern about the implications of increasing monoculture farming practices for biodiversity in the surrounding ecosystems. This coalition of NGOs went as far as to publicly and formally oppose the use of biofuels at the Conference of the Parties of the Framework Convention on Climate Change (Salon.com 2006). There is much debate on these issues, and specific threats and benefits for local communities are extremely difficult to predict with accuracy.

If we (1) truly agree as an international community that there is a global water crisis, (2) are capable of calculating the virtual water flows of agricultural crops, and (3) want to maximize our ability as a global community to produce low-carbon emission fuels, then such an undertaking is imperative.

10.6 The Role Virtual Water Can Play in International Biofuels Policy

As trade in biofuels increases, virtual water metrics should be used to improve global water governance (Fig. 10.4). Achieving integration in governance systems at the global level is a daunting task, but one that is necessary to ensure sustainable and equitable futures as we move toward the use and development of alternative energy sources. Changes in environmental and social institutions are occurring more and more rapidly and it is difficult to reliably track, and thereby comprehensibly understand, system networks and relationships in the current era of globalization (Young et al. 2006). New approaches are needed and it would serve the global community well to think creatively and inventively about the kinds of governing institutions and policies that would be capable of addressing these linked water scarcity problems to causes and solutions. The increasing complexity of networks at the global level challenges us to make sense out of what, at times, could arguably be characterized as managed chaos.

Thus far, policy proposals have addressed virtual water balances and biofuels trade in isolation. Considering these issues simultaneously could lead to three types of policy proposals: (1) domestic regulation, (2) international regulation, and (3) self-regulation within the industry.

(1) A straightforward solution to tackling water-related concerns in biofuel production is regulation at the domestic level. For example, countries may be able to limit irrigation licenses depending on water availability or to set regulative priorities for food over biofuel production when necessary. This will depend in part on a government's ability to collaborate with often powerful agricultural lobbies. Moreover, in many cases the capacity of a single nation-state to handle water issues is limited when water crosses national frontiers. More than 30% of all states meet 30% or more of their domestic water needs from sources in neighboring countries (Wolf et al. 2005, p. 83). Therefore, international agreements are likely needed in many cases in order to achieve sustainable solutions.

(2) A second option would be international regulation which could either mean modification of current international regimes, especially world trade rules, or developing a specific international water regime able to address biofuel trade. In the former case, an obvious option is to integrate biofuel trading into existing international trade regimes in an attempt to harmonize its regulation with other agricultural and energy-related commodities. Biofuels must first be formally classified within the General Agreement on Tariffs and Trade (GATT) framework. In an assessment of the application of World Trade Organization (WTO) and the GATT frameworks to trade in biofuel feedstocks, the International Policy Council identified the lack of commodity classification of biofuels as a significant barrier to incorporation of biofuels into the GATT (IPC 2006).

However, current trade regimes such as the GATT only consider standards which refer to the traded products themselves, without standards regulating the production process. For example, imports of bottled water can be rejected if the quality of the bottled water implies health threats, but there is no way to ban water imported from ground water aquifers which are threatened in terms of water over-use. Modifying current trade rules to internationalize such environmental and social costs related to water use for biofuel production could improve the sustainability of some water systems being used for biofuel production. Other impediments and opportunities include addressing the role of domestic subsidies to biofuel feedstock industries, unilateral and bilateral trade agreement options, and the application of agricultural and environmental standards in trade agreements. One possibility is that the classification of biofuels could be contingent on the fossil fuel-based energy intensiveness, or water intensiveness, of its production if a country is importing the product in an effort to reduce fossil fuel use or CO_2 emissions or water conservation, respectively (IPC 2006).

While classification is important, the development of standards and numerical criteria will play a key role in determining not only where within the GATT biofuels are placed, but how water-related criteria are able to be applied and how subsidy schedules will need to be adjusted within and between countries. This is an area where basic and applied research can significantly improve the global community's ability to address the global trade of biofuels from a virtual water perspective because the potential trade flows will be more predictable once standards allow classification (IPC 2006).

Ensuring that trade in biofuels proceeds in a way that maximizes benefits for those who need them most is critical. The World Food Programme, a United Nations branch organization dedicated to ensuring food security worldwide, realizes the potential that trade in agricultural commodities can have for these communities. Their support of trade as a tool for development is not without reservations, as the achievement of key goals is critical if trade regimes are to benefit a greater number of countries than currently. As stated in their 2002 report, *Reducing poverty and hunger: the critical role of financing for food, agriculture and rural development*, "While there are potential gains from freer trade in farm products, the actual progress made in the ongoing negotiations has been limited so far, and the benefits remain modest. If further liberalisation focuses too narrowly on a removal of OECD subsidies, the lion's share of gains will accrue to developed country consumers and taxpayers. More important for developing countries are: a removal of trade barriers for products in which they have a comparative advantage and a reduction or reversal of tariff escalation for processed commodities; more and deeper preferential access for the poorest of the least developed countries; open borders for long-term foreign investments (FDI); and improved quality assurance and food safety programmes that enable developing countries to compete more efficiently in markets abroad. The resources gained by trade liberalisation and reductions in domestic protection could be channeled into additional development funding (WFP 2002)."

An alternative at the international level to modification of existing international rules is to address water distribution problems through a separate multilateral agreement, such as the proposal of an International Virtual Water Trading Council (Bjornlund and McKay 2001). McKay suggests that such a council could play a role in ensuring that the basic nutrition requirements of people are met through securing food imports and the water saved by these imports could subsequently be used to meet their basic water requirements. Such an entity could act as an independent arm of the World Trade Organization and coordinate international aid efforts with its goals (Bjornlund and McKay 2001). This would be a very valuable contribution to efforts to incorporate virtual water accounting in emerging biofuel trade regimes, as the assurance of basic food and water requirements could serve as a foundation for assessing the capacity for biofuel feedstock cultivation in a given country. In addition, international compacts or conventions among countries to agree to reduce "water footprints," (the extent of water use internally and externally in relation to consumption) (Hoekstra and Chapagain 2006), could provide incentives for countries to create water-efficient trade regimes for biofuel stocks. Such a compact could be loosely modeled on previous agreements such as the Kyoto Protocol, which governs countries' emissions of carbon dioxide.

(3) A final proposal is voluntary self-regulation by the biofuel industry through greater attention to corporate social responsibility (CSR). Biofuel companies are establishing volunteer water saving, and possibly cost saving, standards, committing to the application of water-saving irrigation technologies. Among the most advanced initiatives are the Roundtable on Sustainable Palm Oil (RSPO) and the Cramer Commission (Project Group Sustainable Production of Biomass) (von Geibler 2007, p. 5). While the RSPO goes back to an informal meeting initiated by the World Wide Fund for Nature (WWF) with some multinational corporations, the Cramer Commission was launched by Jacqueline Cramer, the Dutch Environmental Minister. Both the RSPO and the Cramer criteria include regulations on active improvement of the quality and quantity of surface and ground water (Cramer Commission 2007, RSPO 2007). These types of models, when applied to biofuel production and trade, could thus provoke a diffusion of sustainable water management measures in this sector – in particular, as EU governments currently discuss if only certified biomass should be subsidized in future (BMU 2008, Cramer Commission 2007, p. 5).

We see these policy proposals as complementary and far from mutually exclusive. Efforts can be made at all levels to begin to incorporate water-related concerns in trade and development policies using virtual water accounting. Common to all solutions addressing sustainable development of biofuels is the need for more information on their global water requirements. Significant information exists on global land use competition between biofuels and food crops currently and in varying future scenarios, as does the quantification of global virtual water flows within and between regions due to trade in food crops and livestock (Zimmer and Renault 2003, SIWI 2007). The necessity of changing the global distribution of freshwater resources to meet the needs of the poorest people and endangered aquatic ecosystems is also well established (Gleick 1998, Postel 2006). However, many of these data are currently reported at a highly aggregated, global level and are not directly applicable to particular countries. Additionally, sufficient data on the trade (real or potential) in biofuel crops is not available as it is for food crops (IPC 2006). Evaluating these areas is an important first step for all of the sectors and interests involved: those hoping to mainstream biofuel trade, virtually re-distribute water, and secure resources for impoverished people and ecosystems. Such an effort should come from the global community to fulfill its own commitments and interests. However, academic and non-governmental organizations also have a significant role to play in securing the data and interpreting it to promote sustainable global water use.

10.7 Conclusion

As Tony Allan asks, is virtual water a useful concept or a misleading metaphor (Allan 2003)? We think virtual water is a useful way to think about, and empirically assess, the global chain of events that contribute to particular patterns of water use and, subsequently, scarcity and its implications for the production of biofuels. Additionally, the concept can highlight unexplored notions such as "virtual pollution," i.e. environmental burden shift from consuming to producing countries or regions, for example from Europe to Africa. As mentioned earlier, establishing causal links in a global chain is difficult to do through thought experiments, let alone as an empirical exercise. However, virtual water accounting has been developed as a quantitative tool that can and should be used as a criterion for trade and development regimes.

Such quantification is not the end of the story. As Hoekstra and Hung (2005) observe, virtual water can give us the "what," of global water distribution, but it is still up to us to discover the "why." Underlying political and economic conditions may or may not be persuaded by virtual water methods. Incorporating these values into decision making is a worthy goal, however, and should be looked at as an opportunity for future problem solving. It would also be valuable to evaluate not only the virtual water content of agricultural products but the whole production chain for biofuels and how virtual water flows and distribution may coincide with other resource use such as "virtual land" or "virtual timber." Ultimately, it should be recognized that there are externalities to most of the actions we take as individuals or collectively within market economies, and as we continue to develop new energy sources such as biofuels, internalizing these at the global level is a difficult but important task. As we work toward increased food and water security around the world, virtual water accounting will allow us deeper insight and can contribute to more informed decision making.

References

Aldayai MM, Hoekstra AY, Allan T (2008) Strategic importance of green water in international crop trade. In: Value of Water Research Report Series 25, Delft

Allan T (1998) Watersheds and problemsheds: Explaining the absence of armed conflict over water in the Middle East. Middle East Rev Int Aff 2(1):49–51

Allan T (2003) Virtual water – The water, food and trade nexus: Useful concept or misleading metaphor? Water Int 28(1):106–113

Baron JS, LeRoy Poff N, Angermeier PL, Dahm CN, Gleick PH, Hairston NG, Jackson RB, Johnston CA, Richter BD, Steinman AD (2002) Meeting ecological and social needs for freshwater. Ecol Appl 12(5):1247–1260

Berndes G, (2002) Bioenergy and water – The implications of large-scale bioenergy production for water use and supply Global Environ Change 12:253–271

Biswas A.K. 2004. Integrated water resource management: A reassessment. Water Int 29(2): 248–256

Bjornlund H, McKay J (2001) Elements of an institutional framework for the management of water for poverty reduction in developing countries. In: Unver IHO, et al. (eds) Water development and poverty reduction. Kluwer Academic Publishers, Norwell, MA

BMU (2008) Nur nachhaltig angebautes Palmöl kann Beitrag zum Klimaschutz leisten. Studie zu den Folgen der Palmöl-Nutzung vorgelegt. In: Federal Ministry for the Environment, Nature Conservation, Nuclear Savety (ed), News Service, no. 026/08, Berlin, 21 February

Bringezu S, et al. (2007) Towards a sustainable biomass strategy. Wuppertal Discussion Paper no. 163, Wuppertal

Chapagain AK, Hoekstra AJ (2004) Water footprints of nations, vols. 1 and 2, Main Report. UNESCO Institute for Water Education, Delft

Conca K (2006) Governing water: Contentious transnational politics and global institutions building. MIT Press, Cambridge, MA.

Cramer Commission (2007) Testing framework for sustainable biomass. Final report from the project group "Sustainable production of biomass," February 2007, http://www.lowcvp.org.uk/assets/reports/070427-Cramer-FinalReport_EN.pdf

Dregne HE (1986) Desertification of arid lands. In: El-Baz F, Hassan MHA (eds) Physics of desertification. Martinus, Nijhoff, Dordrecht, The Netherlands

von Geibler J (2007) Biomassezertifizierung unter Wachstumsdruck. Wie wirksam sind Nachhaltigkeitsstandards bei steigender Nachfrage? Diskussion am Beispiel der Wertschöpfungskette Palmöl, Wuppertal Paper 168/2007, Wuppertal

Gleick PH (1998) Water in crisis: Paths to sustainable use. Ecol Appl 8(3):571–579

Global Forest Coalition et al. (2006) Biofuels: A disaster in the making. www.energybulletin.net

The Global Water System (2005) Project Science framework and implementation activities. Earth System Science Partnership

Gonsalves JB (2006) An assessment of the biofuels industry in India. United Nations conference on trade and development, 18 October 2006

Hazell P, Pachauri RK (2006) Bioenergy and agriculture: Promises and challenges. 2020 Focus 14. International Food Policy Research Institute, Washington, DC

Hoekstra AY, Hung PQ (2005) Globalisation of water resources: international virtual water flows in relation to crop trade. Global Environ Change 15:45–56

IRIN News (2006) South Africa: Ethanol – Boon or bust?

Hoekstra AY, Chapagain AK (2006) Water footprints of nations: Water use by people as a function of their consumption patterns. Water Resour Manage 60:1

International Policy Council (IPC) (2006) WTO disciplines and biofuels: Opportunities and constraints in the creation of a global marketplace. IPC Discussion Paper

International Water Management Institute (2007) Water for food, water for life: A comprehensive assessment of water management in agriculture. Earthscan, London; International Water Management Institute, Colombo

Johnson N, Revenga C, Echeverria J (2001) Managing water for people and nature. Science 292(5519):1071–1072

Jungbluth R, Rohwetter M (2007) Wassermangel. Raubbau am kostbarsten Gut. In: Die Zeit 15:25, http://www.zeit.de/2007/15/Nestle-Interview-Brabeck

OECD Observer (March 2006). Virtual Solution. http://www.oecdobserver.org/news/fullstory.php/aid/1798/Virtual_solution.html

Postel S (2006) Safeguarding freshwater ecosystems. In: Worldwatch Institute (ed) State of the World 2006: Special focus: China and India. W.W. Norton and Company, New York, pp. 41–76

Rockstrom J, Axberg GN, Falkenmark M, Lannerstad M, Rosemarin A, Caldwell I, Arvidson A, Nordstrom M (2005) Sustainable pathways to achieving the millennium development goals: Assessing the key roles of water, energy and sanitation. Stockholm Environment Institute, Sweden

RSPO (2007) RSPO principles and criteria for sustainable palm oil production including indicators and guidance. www.rspo.org/resource_centre/RSPO%20Principles%20&%20Criteria%20Document.pdf

Salman MA (1999) Groundwater: Legal and policy perspectives. World Bank Technical Paper 456, Washington, DC

Salon.com. (2006) Biofuel neocolonialism? Senegal wants to be part of a green OPEC, India and Brazil want to help. www.salon.com/tech/htww/2006/11/01/senegal_biofuel.html

Schipper L, Marie-Lilliu C (1999) Transportation and CO_2 emissions: Flexing the link – A path for the World Bank. In: International Energy Agency Environmental Department Paper 69, Paris, France

SIWI (2007) Scenarios on economic growth and resource demand. Background report to the Swedish Environmental Council 2007

Spiegel Online (2007) Spaghetti shock in Italy, biofuels boom results in pricey pasta, 07/11/2007. www.spiegel.de/international/business/0,1518,druck-493795,00.html

Turton A, Nicol A, Earle A, Meissner R, Mendelson S, Quaison E (2003) Policy options in water-stressed states: Emerging lessons from the Middle East and Southern Africa. African Water Research Unit (AWIRU) and Overseas Development Institute (ODI), Pretoria, London

United Nations Conference on Trade and Development (UNCTAD) (2006) The emerging biofuels market: Regulatory, trade and development implications. United Nations, New York, 48p

UNDP (2006) Human development report 2006. Beyond scarcity: Power, poverty and the global water crisis. United Nations Development Programme, New York

UNEP (2001) The water crisis. United Nations Press, New York

UNEP (2006) Challenges to international waters – Regional assessments in a global perspective. United Nations Environment Programme, Nairobi, Kenya

UNESCO (2006) Water a shared responsibility. The UN World Water Development Report 2, World Water Assessment Programme. Berghahn Books, New York, Paris

Unver I H, Olcay R, Gupta K, Kibaroglu A (2003) Water development and poverty reduction. Kluwer Academic Publishers, Norwell, MA

WSSCC (Water Supply and Sanitation Collaborative Council) (2000) Vision 21: Water for people – A shared vision for hygiene, sanitation, and water supply and a framework for action. WSSCC, Geneva

Wolf AT, Kramer A, Carius A, Dabelko GD (2005) Managing water conflict and cooperation. In: Worldwatch Institute (ed) State of the World 2005. Redefining global security. W.W. Norton & Company, New York, pp. 80–95

World Food Programme (2002) Reducing poverty and hunger: The critical role of financing for food, agriculture and rural development. Prepared for International Conference on Financing and for Development

World Resources Institute (WRI) (2000) World Resources 2000–2001, people and ecosystems: The fraying web of life. World Resources Institute, Washington, DC

Young OY (2002) The institutional dimensions of environmental change: Fit, interplay and scale. MIT Press, Cambridge, MA.

Young OY, Berkhout F, Galloin GC, Janssen MA, Ostrom E, van der Leeuw S (2006) The globalization of socio-ecological systems: An agenda for scientific research. Global Environ Change 16:304–316

Ziegler J (2008) Report of the special rapporteur on the right to food, UN Human Rights Council, GE.08-10098 (E) 180108, www.humanrights.ch/home/upload/pdf/080311_bericht_ziegler.pdf

Zimmer D, Renault D (2003) Virtual water in food production and global trade: Review of methodological issues and preliminary results. World Water Council Report: http://worldwatercouncil.org/fileadmin/wwc/Programs/Virtual_Water/VirtualWater_article_DZDR.pdf

Chapter 11
Whose Nature – Whose Water? Some Remarks About the History of Ideas, Property and Democracy of Water

Uta von Winterfeld

> *One of the important questions in pre-modern times was: How to compose "property" – if God gave the world, gave nature to all his children, to mankind in common?*

Abstract First of all this article shows how our comprehension of nature is still influenced by Francis Bacon. Then the author discusses the concept of property as developed by John Locke, which still influences neoliberal theories, followed by a discussion of the concept of the commons and conceivable revisions to it. The article criticizes the privatization of common goods like water and makes claims for sustainable and fair handling of global water resources.

11.1 Introduction

In the dawn of the Age of Enlightenment and in addition to the purely geographic journeys of conquest and Eurocentric world trade, yet another dimension came into being.[1] It was the English Lord Chancellor Francis Bacon who at that time, in his work entitled "Novum Organum", presented the notion that in the course of the general invention of government, nature would also have to be governed. He wanted this to be understood as a kind of journey of intellectual conquest by way of which, now that the New World had been discovered physically, it should also and ultimately be discovered intellectually. Among the many postulates in his programme for the dominance of nature, Francis Bacon determined that:

> ... our principal object is to make nature subservient to the state and wants of man, ... (Adler 1996, p. 159).

U. von Winterfeld (✉)
Wuppertal Institut für Klima, Umwelt, Energie, Wuppertal, Germany
e-mail: uta.winterfeld@wupperinst.org

[1] See Claudia von Braunmühl and Uta von Winterfeld (2005), Wuppertal Paper No. 135e, March 2005, pp. 26–28 (transl: Stewart Lindemann).

"... nature is only subdued by submission, ..." (ibid., p. 137). This means that one can subdue nature only by submitting to her; the intellect will have to adapt itself to nature if it wishes to dictate her actions.

With the benefit of divine grace he, Francis Bacon, decorated the nuptial bed of intellect and nature. "The wedding song's entreaty is that the issue of this marriage may be assistance to humankind and a line of inventors, to ease and banish need and desolation".

In this "Program for government", aimed at domination, nature is firstly thought of as being homogenous, like the human states and wants. All across the face of the earth nature is to be made subservient to human purposes, which are everywhere identical. Secondly, nature serves exclusively human ends. She has no other, let alone her own, agenda. Thirdly, Francis Bacon conceptualizes a "clever" dominance of nature, one which is adapted to nature. The mechanical arts of his time – today referred to as technology – are to dominate nature in accordance with and not contrary to nature. Fourthly, nature is granted no realm of action on her own; she is to develop in accordance with human purposes. Fifthly, "nature" is thought of as being feminine whilst "intellect" is masculine; but they are to be reconciled by marriage one with the other. There derives from this, sixthly, a masculine lineage of inventors with whose aid the vicissitudes of the world would best be mastered (Fig. 11.1).

Fig. 11.1 Natura. Copperplate of Martin Heemswerk, about 1572

Many of Bacon's comments are appropriate to his own age. What remains for us today however, is Bacon's programme: By mastering nature it is possible to return to Paradise on earth and to live in prosperity.

The necessary premise of mastering nature is to own her. And the necessary premise of ownership is to objectify the other as a thing and a resource. Against this background, I will first present some ideas of John Locke about private property, secondly show the problematic nature of these patterns of rationalism relating to the commons, and thirdly reflect on water conflicts and principles of water democracy.

11.2 John Locke – His Ideas About Nature, Labour and Private Property

The final purpose of civil society is, so the statement of John Locke in his second Treaty of Government, the preservation of property. If a government is not able to protect property, it can be deposed. But how does property come into being? John Locke derives it from the "State of Nature" in his fifth chapter. Within this state of nature everything belongs to all in common because God entrusted it to all human beings. But then...

> Though the earth and all inferior creatures be common to all men, yet every man has a 'property' in his own 'person'. This nobody has any right to but himself. The 'labour' of his body and the 'work' of his hands, we may say, are properly his. Whatsoever, then, he removes out of the state that Nature hath provided and left it in, he hath mixed his labour with it, and joined to it something that is his own, and thereby makes it his property. It being by him removed from the common state Nature placed it in, it hath by this labour something annexed to it that excludes the common right of other men. For this 'labour' being the unquestionable property of the labourer, no man but he can have a right to what is once joined to, at least where there is enough, and as good left in common for others. (Locke 1823/1690, p. 116)

The last subordinate clause shows an interesting bound: We can only have private property if there are enough – and as good – of commons. Another thing we read is that John Locke tells us a story of men, of every man and *his* labour. You may argue that this is only a problem of language. But it is probably not, because Locke tells us the story of Adam and his sons. And the truth behind it is: The workman of John Locke has two properties: The fruits of his work and the – fruitless – work of his wife. But this is another story, analyzed very well by Carole Pateman within her book "The Sexual Contract" (1988). Let us continue with John Locke and how he explains property coming into being and natural things coming into value.

The labourer is nourished by the acorns he picked up under an oak and by the apples he gathered from the trees in the wood. This labourer has certainly appropriated the fruits to himself. His labour put a distinction between apples, acorns and common. That labour adds something which is more than Nature, "the common mother of all" (ibid.) had done. So they became his private right. Corresponding John Locke continues:

> Though the water running in the fountain be every one's, yet who can doubt but that in the pitcher is his only who drew it out? His labour hath taken it out of the hands of Nature where it was common and belonged equally to all her children, and hath thereby appropriated it to himself. (ibid., p. 117)

Following John Locke, this is a law of reason which also makes the deer that the Indian has killed his own. But there is another bound of property:

> The same law of Nature that does by this means give us property, does also bound that property too... As much as any one can make use of to any advantage of life before it spoils, so much he may by his labour fix a property in ... Nothing was made by God for man to spoil or destroy. (ibid., p. 117 and 118)

Unfortunately this important bound only works without money – because money doesn't spoil. According to John Locke, gold, silver and diamonds are things that agreement has put into value. Without this agreement, they are useless to the life of man.

The chief matter of property is, as John Locke sees it, earth itself – and it is like the picked acorns and gathered fruits, like the water in the pitcher and the killed deer:

> As much land as a man tills, plants, improves, cultivates, and can use the product of, so much is his property. He by his labour does, as it were, enclose it from the common. (ibid., p. 118)

What John Locke outlines here is a first draft of a theory of the value of labour. What does put things into value? And what about the relationship between labour and nature? The author continues:

> I think it will be but a very modest computation to say, that of the products of the earth useful to the life of man, nine-tenth are the effects of labour. Nay, if we will rightly estimate things as they come to our use, and cast up the several expenses about them – what in them is purely owing to Nature and what to labour – we shall find that in most of them ninety-nine hundredths are wholly to be put on the account of labour. (ibid., p. 122)

What an amazing story! Nature loses 90 hundredths of her value within two sentences. To sum it up it can be said that the transformation from common nature in private property takes place through labour. It is combined with special concepts of biblical order and progress: Subduing or cultivating the earth and dominating it are joined together. The common nature is wild, uncultivated and useless. Progress derives from human labour which transforms nature into private property.

But what about the first found bound – private property is only possible if there are enough commons?

11.3 The Tragedy of the Commons and Water as a Global Public Good

The debate on *commons* was instigated by Garrett Hardin (1968) and immediately became linked with the term "tragedy". But reviews of his essay on the "Tragedy of the Commons" have frequently overlooked the fact that Hardin was actually focussing on the problem of population growth. He was searching for moral arguments for the restriction of liberties, or for achieving moderation – both in the use of *commons* as well as in reproduction. According to my view, this combination is highly problematical, but is hardly discussed.

If a common good is left for free use, then the tragedy for Garrett Hardin is that everybody tries to use as much of the limited resource as possible for their own advantage, which finally leads to the ruin of all (Hardin 1968, p. 1244). The situation is similar with the problem of pollution. Something which is available to all is exploited and defiled as long as all behave as independent and rational free entrepreneurs (ibid., p. 1245). His conclusion is not the limitation of free entrepreneurship, but the abolition of the communal exploitation of commons. The system of private property and its inheritance is unjust but such an injustice is preferable to the ruin of all (ibid., p. 1247). In a later text (Hardin 1978) he sees only two possible solutions: private enterprise or socialism (control of the commons by the government). He does not see other options.

The best known response to Garrett Hardin has been formulated by David Feeny, Fikret Berkes, Bonnie J. McCay and James M. Acheson. They argue expressly against setting up a stark alternative between Market and State. In "The Tragedy of the Commons: Twenty-two Years Later" (Feeny et al. 1990) they also tackle the assumption that the communal use of resources leads to their exploitation or degradation. Their first step is to distinguish between the resource itself and the ownership rights. Instead of commons they use the term "common-pool resources". What Hardin refers to as a tragedy is actually the lack of well-defined ownership rights. This applies for "open access" elements, which are subject to a free-for-all. But it does not apply in the case of communal property, which is managed by an identifiable community where the users are dependent on one another. The authors distinguish between four categories of ownership rights: free access, private ownership, common ownership and state ownership. For the latter three categories, they present both positive and negative examples of sustainable and non-sustainable use of common goods. Therefore, it is not possible to say for any form of property that it was the only one or the best for a sustainable approach to commons. But what Garrett Hardin referred to as the tragedy of the commons only applies after conditions have been created which allow free access to common goods. However, this was usually the result of the introduction of colonial rule, for example in the sub-Saran region or on the Pacific Islands (ibid., p. 6).

Correspondingly Elinor Ostrom et al. formulate a concept of commons which indicates that the use of common goods is inseparably linked to the question of property rights. "When no property rights define who can use a common pool resource and how its uses are regulated, a common pool resource is under an open-access regime" (Ostrom et al. 2002, p. 18).

Entering into my analysis of the concept of the composition of private property by John Locke I want to formulate another dimension of the tragedy of the commons. Analogous to the workman who has two properties, the fruits of his work and the – fruitless – work of his wife, private property itself has two sources, labour

Fig. 11.2 Wildlife Park in the West of Chile, Chile

and/or money and an endless pool of commons. Valuable labour needs valueless labour and valuable private property needs valueless natural commons (Fig. 11.2).

The tragedy here is that valuable labour needs to exploit a hidden resource and that private property tends to destroy natural commons. Both are the contrary of sustainability; it causes a crisis of social and ecological reproduction and regeneration and endangers the base of productivity. This problem becomes still more serious with globalization.

The threat to or neglect of public duties and public goods, according to the Final Report of the Enquete Commission "Globalization of the World Economy – Challenges and Responses", represents perhaps the most significant threat of a globalization which concentrates mainly on the increase in private goods by increasing global efficiency (Enquete 2002, p. 56). The Study Commission is oriented towards a "socially and environmentally sustainable conservation, protection and management of public goods" (ibid., p. 418). It obviously hopes to be able to counter this effectively with the concept of global public goods. These public goods include peace, respect for human rights, social justice and an "intact environment", for example "Climate" is a global public good (ibid., p. 56). To answer the question whether "global public goods" are a solution or part of the problems combined with the domination of private property and private goods, I will look at this approach, and in particular the work of Inge Kaul et al.[2]

In recent years the concept of global public goods has established itself as a new frame or reference for the international debate on liberalization and privatization, the public or private nature of goods, and the future role of states and international

---

[2]Ibid., p. 29f, translated by Richard Holmes.

organizations. The discussion about the contours of the concept and the strategies for financing and managing global public goods are still in their early stages. When it was adopted in 1998/1999 in UNDP circles (see Kaul et al. 1999), this was against the background of the financial crisis in Asia and repeatedly postponed conference on development financing. The central problems were the bottlenecks in development cooperation and the attempt to reinterpret (reframe) requirements and impacts as a way of mobilizing more resources from the private economy for development cooperation. Development facts, the argument runs, should be weighted with respect to reciprocity and collective responsibility (ibid., p. xiii) and to this extent should be viewed in terms of an understanding of global public goods. The book published in 1999 has the sub-title "International cooperation in the twenty-first century" and redefines a whole range of sectors from road signs through biodiversity to peace and just distribution in the light of the new frame of reference.

The impulse met with considerable interest, although the debate has not yet produced consolidated results or led to conceptual clarification. Analytical and normative dimensions of the concept of global public goods are often unwittingly mixed. However, positions are beginning to crystallize out. In particular civil society proponents tend to group around the concept of commons or common goods, such as air and water, which are not commodities in any form and which should not be subjected to economic exigencies as traded goods (e.g., Petrella 2000, Shiva 2002).

Efforts from the civil society into the field of political institutions in order to link up to the concept of global public goods and to give this a form which mixes public and private elements specifically for each situation. But there the public character of a good is not based on the property of the good itself, but is regarded as a social construct and political convention. A good becomes public or private only as a result of regulatory measures. These measures are disputed and are the subject of social and political conflicts. Every definition of global public goods and each decision about what should be treated as such therefore represents a political act, which must be subject to decisions within democratic structures and procedures.

The privatization of an area of supply which was previously recognized as a public good is always associated with a loss of public openness and thus with a restriction of democratic participation. The context of globalization, deregulation and privatization and the extreme pressure in particular regarding water management in the North and South, not least from the powerful international development agencies towards the privatization of water supplies, has contributed considerably towards making public goods a topic of debate. Is the concept of global public goods suited to preserving existential public services, an orientation to public welfare, and social controllability as values and the social practice? Concerning the neoliberal pressure on water as a resource and as a freely tradable good, which attitudes and actions are most suited and able to oppose this pressure?

Inge Kaul, Pedro Conceicao, Katell Lo Goulven, and Ronald U. Mendoza tackled this problem in their book "Providing Global Public Goods: Managing Globalization" (2003) with a clear emphasis on the level of implementation. Globalization, according to the book, gives rise to global goods as well as global bads, and at the same time is driven by these (Kaul et al. 2003, p. 3). They are therefore an element of globalization. If this is to be managed, then the relationship between the public and private must be re-invented. Concepts of public and private are in most cases social constructs. Public goods are usually described as goods of the public realm, but not goods provided by the state (ibid., p. 7). State authorities and markets can themselves be regarded as public goods (ibid.). Here things become unclear and it sounds almost as if the problems associated with globalization can be solved in a new public–private partnership by self-management of the global goods "state" and "market".

In a historical review of public goods (ibid., pp. 63–77) it is emphasized that public goods are often thought to be provided by the state. This, according to the authors is false in the current situation, because non-state actors are increasingly involved at both national and international levels. In addition, state expenditure for public goods is relatively young, historically speaking. Until the seventeenth century most state funds were spent on war. It was only during the course of industrialization that countries began to spend a significant part of the state budget on the provision of public goods such as water. Often this did not happen without struggle. A theory of public goods was developed in the "Golden Years" of Keynesianism from 1945 to 1975, which at the same time was marked by enormous state growth. However, this period was an exception. Therefore, it is necessary to take a new look at public goods and revise the concept. Public goods are not simply provided at the global level, and the public must be involved much more in the formulation of preferences. Participation is not required in the consumption of public goods, but in their production. How this is to be achieved remains an open question.

Global public goods, according to Kaul et al., can be regarded as the sum of national public goods and international cooperation. Therefore, the largest part of global public goods is nationally financed (ibid., p. 36). The financing of global public goods (such as international communication or transport systems) often functions very well and attracts private investors, because charges can be levied. However, there are many other global goods that are dependent on subsidies, as witnessed by the fact currently some 30% of development aid is in the form of global public goods (ibid., p. 38). In the vision of the international community within the international public sphere ten global goods are identified, namely:

1. basic human dignity
2. national sovereignty
3. global public health
4. global security
5. global peace
6. harmonized, trans-boundary communication and transport systems
7. harmonized, trans-boundary institutional infrastructures
8. common management of knowledge
9. common management of global natural "commons" in order to promote their sustainable use
10. the availability of international arenas for multilateral negotiations between states, and also between states and non-state actors (ibid., p. 44)

This shows, firstly, that "global public goods" refers to very different things at different levels. Secondly, another concept is used when talking of sustainability, namely not "goods" but "commons". Correspondingly, water could be included twice in this list, under institutional infrastructures and global natural "commons".

The concept of global public goods is only in part able to set a counterpoint to the mainstream of concentration on the increase of private goods. The intended new relationship between public and private as well as state and market obscures the fact that to this day both have partially failed with respect to global natural commons. The new nexus of market and state can at best be applied to the supply and disposal, i.e. to the infrastructure but not to water as a common. But basically infrastructures and commons point to a similar problem: Whereas the immense intensification of water management is intended to market the resource water as profitably as possible, not only the need for its regeneration has been neglected but also the ecological water cycle.[3] Whereas elements of infrastructure are currently being removed from the scope of public services and made freely tradable commodities, there is a risk that the specific local environmental requirements for an appropriate and sustainable water and sewerage service will be ignored, as will the social context of the infrastructures. Against this background there is not really a need for a conception of global public goods, which re-configures the public and the private spheres. Rather, in connection with sustainable governance there is a need for a political approach, which encourages and protects a varied sustainable governance from below for each locality and is active for this in an international context.

11.4 Water Conflicts and Principles of Water Democracy

Scarcity of and also greed for resources are the main reasons for conflicts and wars. The story of water wars and water in conflict is a big and long story, which comprises different levels. Water has for example been said to be the next oil, as the looming source of world conflict, as more valuable than gold or within the context of the dark side of natural resources. The Pacific Institute for Studies in Development, Environment and Security sees Water rarely, if ever, as the only source of violent conflict, but emphasizes that there is in fact a long history of conflicts and tensions over water resources. The Institute initiated a project on the history of water conflict and elaborated a chronology (since 1503) as well as a Water Conflict Bibliography. The Global Policy Forum (monitoring policy making at the United Nations) showed and elaborated an analysis and numerous articles and documents about water in conflict. Inside this broad debate I want to point out three aspects: Dams and water conflicts between nations, water conflicts and livelihood, and climate change and the water crisis.

---

[3]Significantly, the first chapter of the Gender and Water Development Report (2003) is dedicated to the topic of "Water for Nature".

Until today, water conflicts reflect the programme of Francis Bacon and the paradigm about property of John Locke: By mastering nature it is possible to return to Paradise on earth and to live in prosperity – and progress derives from human labour which transforms nature into private property. For European colonizers who came to America, river colonization was a cultural obsession and an imperial imperative (Shiva 2002, p. 53).

At the beginning of the twentieth century, WJ McGee, the chief adviser on water programmes of the American president Theodore Roosevelt (1901–1909) stated that water was "the single step remaining to be taken before Man becomes master over Nature".[4] And in 1928, an irrigation scientist with the Bureau of Reclamation, John Widtsoe, argued:

> The destiny of man is to possess the whole earth; and the destiny of the earth is to be subject to man. There can be no full conquest of the earth, and no real satisfaction to humanity, if larges portions of the earth remain beyond his highest control. Only as all parts of the earth are developed according to the best existing knowledge, and brought under human control, can man be said to possess the earth.[5]

Rivers managed by the community and following their ecological path were viewed as wasteful. One central measure of states taken to control water resources and to collaborate with private entrepreneurs is: building dams. Also in the field of development policy dams have been seen as a sign of development and giant water-project loans from the World Bank enabled governments to control water resources and to colonize rivers and people.

The annual consequence is a global picture of displacement. According to the World Commission on Dams 40–80 million people have been displaced by dam projects – and it may be doubtful whether this phenomenon is linked with social and ecological – or purely with economical benefits for the constructing enterprises. The commission concludes that too often "an unacceptable and often unnecessary price has been paid to secure those benefits, especially in social and environmental terms, by people displaced, by communities downstream, by taxpayers, and by the natural environment".[6]

Large dams also modify the distribution patterns of water and shift water allocation. This – on many occasions – generates interstate water conflicts between nations. In 1961, Mexico protested that water flowing from the United States was heavily restricted by dams and until today the United States and Mexico are in conflict over Colorado River waters. In 1974, water development projects were the cause of armed conflict between Turkey, Syria, Iraq and the Kurds over Euphrates river waters. And in 1989, the Turkish prime minister Turgut Ozal threatened to use water against the militants by cutting supplies off entirely unless Syria expelled the PKK, to whom it had given refuge (Shiva 2002, p. 72).

---

[4] See Goldman et al. (1973).

[5] John Widtsoe, "Success on Irrigation Projects", published as a pamphlet in 1928, p. 138.

[6] Dams and Development (2000) Report of the World Commission on dams. Earthscan Publications, London, p. xviii.

To some extent, the war between Israelis and Palestinians is a war over water. While only 3% of the Jordan basin lies in Israel, the river provides 60% of its water needs. In 2000, 50% of the total cultivated area in Israel was irrigated. Palestinian villages consumed only 2% of Israel's water. The water apartheid is fuelling the already heated Israeli–Palestinian conflict (ibid., p. 73).

The story of conflict over the Nile – the longest river in the world, shared by ten African countries with an estimated total population at 850 million – is also a story about amazing agreements. In 1903 the British (who used the Nile in the Sudan for navigation) signed an agreement with Ethiopia not to manipulate the flow of the Blue Nile. But in 1958, Egypt began building the Aswan Dam and displaced 100,000 Sudanese. In 1959, Egypt and the Sudan entered a bilateral agreement regarding the full utilization of Nile waters – regardless of the water demands, potential or otherwise, of the upper riparian states (ibid., p. 75).

In 1997, the United Nations wanted to create guidelines for water sharing of international rivers and tried to establish two principles: the rule of equitable and reasonable use (water sharing on an equitable basis among multiple users) and the no harm rule (not causing harm to co-riparian states). In 1999, the ten Nile basin states endorsed a Nile River Basin Strategic Action Program and tried to go beyond past conflicts and sustainably and justly use the waters of the world's biggest river for some of the world's poorest people (ibid., p. 76).

This first aspect, dams and water conflicts between nations, shows that water conflicts have to be seen also as conflicts of power – USA, Turkey, Israel or Egypt as stronger nations restrict the access to water of Mexico, Iraq, Palestine or Sudan as weaker or non-nations.

The second aspect, water conflicts and livelihood, shows another dimension of water and power conflicts. For many people, water is still a "critical resource" that determines the success of a subsistence economy and that may endanger survival. I don't want to discuss here the conflict between the capitalist greed for profit and the sustenance need of indigenous people but I do want to show something of the quality of water conflicts and water management by people whose economy is still significantly dependent on nature and water.

Eileen K. Omosa[7] elaborated a case study of the Wajir District in Kenya and examined the impact of water conflicts on pastoral livelihood. The district lies in the north-eastern part of Kenya and is a water-scarce region. Annual rainfall is unreliable and between 200 and 300 mm. It has a population of between 300,000 and 350,000 people and covers an area of 56,000 km^2. Part of the data of this report was obtained from primary sources and collected by face-to-face interviews with a sample of 100 household heads.

Pastoralists are livestock herders and found throughout Africa's arid regions. Wajir District is 100% arid and semi-arid land, with rangelands suitable for pastoralism, and with small parts suitable for annual crop production. It is characterized

---

[7] August 2005, International Institute for Sustainable Development (iisd). Pdf: www.iisd.org/pdf/2005/security_pastoral_water_impacts.pdf

by clan settlements. The community of the four clans is patriarchal. The large number of male-headed households implies that men make decisions on access to water and management of arising conflicts. Both tend to be inclined towards male needs. Eileen K. Omosa emphasizes the sad side of water conflicts: Most victims are civilians; mainly women and children (Osoma 2005, p. 4).

Within Wajir District, availability of water determines where people and livestock settle during a given year. During the dry season, people rely solely on boreholes because all the temporary water sources dry up. Pastoralism as a livelihood relies on the availability of pastures and labour to thrive with water as the determining factor. It is a highly flexible system with increasing livestock numbers in good seasons that provide subsistence during the dry seasons. The resource-use pattern is characterized by risk-spreading and flexible mechanisms like mobility, communal land ownership or diverse herd sizes.

In the 1950s, pastoralism was losing its hold in the economic, social and political system of Kenya. It was dominated by the needs of export agriculture and did not support pastoralists because they have less to offer to the state, especially in terms of resources for the export markets. Pastoral communities were seen as unwilling and unable to modernize. The government distrusted them, assuming a lack of national loyalty because of their cross-border movements.

Pastoral communities are quickly and existentially affected by conflicts over water. At the household level, 69% felt the impact of conflicts as reduced access to food. Sixty-one percent mention the interruption in education and 59% the reduced health care service. Conflicts over water can mean loss of life and property, degeneration of social relationships or forced migration of families and livestock – serious implications to people "already experiencing figures below the national average in basics like health, education and nutrition" (ibid.). And there is a vicious circle of conflicts over water – insecurity leading to reduced activities outside a secure home – livestock as the main source of milk, meat and income are negatively affected – new conflicts arise. And the time taken in managing conflicts is lost and means a reduction in labour.

Traditionally, a clan had structures that provided the basic framework for accessing water and other natural resources. Digging, use and maintenance of wells were governed by an elaborate system of customary rules and access was clearly understood to be part of reciprocal agreements. But when the government handed over the new and modern boreholes to the community, it had little orientation on how to manage such modern watering points with its associated technologies. Borehole management became a problem. Consequently, OXFAM-GB introduced a concept of community management through Water User Associations (ibid., p. 11), which may be seen as a new beginning of common democratic water management. But experience has shown that people abide by laws and rules as long as the resource is sufficient (and there is equity). Therefore, traditional experiences with and basic framework for accessing water in water-scarce regions may be important for sustainable water politics, which have to harmonize modern and traditional practices of pastoralism.

Water conflicts between nations as well as the impact of water conflicts on livelihoods are and will be influenced by a third aspect – climate change deepening the world water crisis. The main impact of climate crisis on all forms of life is mediated through water in form of floods, cyclones, heat waves and droughts. Vandana Shiva mentions a killer cyclone which hit the eastern part of the state of Orissa in Eastern India in October 1999. The cyclone damaged 1.83 million houses and 1.8 million acres of paddy crops in 12 coastal districts (Shiva 2002, p. 39). Two years later, Orissa was first hit by one of the worst droughts in history and then affected by the worst flood during the monsoon season. The calamities in Orissa are not isolated disasters but part of climate change-related incidents. In 1996, 600 people in Angola were killed by a typhoon and a deadly blizzard in the Western Chinese Highlands pushed 60,000 Tibetan herders to starvation. Twenty of Laos' rice fields were damaged due to flooding and 330 people died in Yemen due to the worst floods in 40 years (ibid., p. 46). But climate change is also aggravating drought and heat waves. In 1995, an area in southern Spain suffered its fourth consecutive year of drought and in June, temperatures in Russia reached 93 degrees Fahrenheit. Three hundred people in India and about 500 people in Chicago were killed by heat waves (ibid., p. 49).

The main victims of climate disasters are coastal communities, people living on small islands, peasants, and pastoral communities – those who have had the smallest role in creating climate destabilization. For example Mangroves: As coastal fringe forests they can survive in the saline wetlands. Their destruction means tremendous problems of erosion and siltation and sometimes enormous loss of human life and property. Climate injustice as water injustice means that the most vulnerable people and regions in the South are affected by the impacts of the most inconsiderate actions of the North. Sunita Nairan, director of the New Delhi-based Centre for Science and Environment mentions the challenge of climate change and declares: "And it is this challenge which the world is completely failing to do anything about, and which will jeopardize the water security of large numbers of people, who already live on the margins of survival".[8]

When the impacts of industrial northern activities threaten the vulnerable south, there is need for action. But Sunita Narain told the Inter Press Service (see footnote 8): "... the international community does not understand water and how it affects local communities and, therefore, the United Nations and the international community is looking for quick fix technological solutions to what is primarily a governance issue".

If the solution of water conflicts is primarily a governance issue, then it is necessary to return to the commons and to consider a sustainable and democratic relationship with water.

---

[8]Cited in: Climate Change Deepening World Water Crises, by Thalif Deen, Interpress Service, March 19, 2008 (Global Policy Forum), www.globalpolicy.org/socecon/envronmt/climate/2008/0319deepwater.htm. Accessed 10 April 2008.

Commons, Vandana Shiva emphasizes, are not resources with open access. They require a concept of property at the level of user groups, which establish rules for their use. Nor is resource management linked to private individuals (Shiva 2002, p. 26). Against this background, the statements of Garrett Hardin about the tragedy of the commons seem factually and historically questionable. This point is raised among others by Gerhard Scherhorn (1998a, b). The displacement of common goods, he explains, should be interpreted somewhat differently than the legend of greedy users. Historically, the common land in Europe was enclosed as grazing land (for wool production), and the local small-holders who had previously been able to use it were driven away. The principle of common goods was not undermined by the users but was broken by the usurpers.

If the breakdown of the commons is not attributed to the search for benefit by individual users, but to a new form of cooperation between state and business (which promotes private property and undermines communal property), then this results in new solution strategies. At the same time these are associated with another approach to politics: In there actions, people are not only (and possibly also not primarily) motivated to be socially and environmentally aware by material advantages (market) or drastic sanctions (state). Rather, they are also able to act with awareness simply if they are allowed to, or when their socially and environmentally motivated actions are not undermined.

However, there is an obvious danger of painting a picture of traditional societies free from all ambivalence, which is also shown by the mentioned example of pastoral livelihoods in Wajir district in Kenya. For Vandana Shiva, for example, it is the indigenous communities who created the rules and limitations for water use and which guarantee sustainability and equality in the dealings with water (Shiva 2002, p. 12). For Gerhard Scherhorn feudal societies were based on a principle of mutual solidarity, in which peasants worked the land owned by their feudal lord, who in turn represented and protected them (Scherhorn 1998b). He does, though, concede that while this was the *ideal*, the reality may often have been rather different (1998a). There is a considerable risk of idealizing traditional societies, particularly when they can serve as a counter-model to current globalization developments. One cannot assume that indigenous or feudal societies were per se more free, more sustainable or more egalitarian. In particular the peasants in the later Middle Ages were feudally subordinate. And the pre-colonial access to water may frequently have been regulated by hierarchies – including gender hierarchies. It is hardly possible to derive principles from this for a sustainable governance in the current situation. However, there is one parallel: Local communal management and communities are broken up in accordance with central or centralistic requirements, which immediately set themselves up as irreplaceable and permit no alternatives.

One alternative path might be an "integrated water management" as proposed by Lyla Metha in connection with the concept of global public goods (see Kaul et al. 2003, pp. 556–575). She queries the use of the term public with regard to access rights to water, and emphasizes that water is far from having the properties of a global public good. The access to water reflects power asymmetries, shows socio-economic inequalities and other distribution factors such as land ownership (ibid.,

p. 556). Therefore, it is important to distinguish between the abstract designation of water as a common resource and water as a resource in real life. Integrated water management could take account of such differences and can be based on the access to water as a human right. The supply of water must be organized on the basis of the various water systems; they must sometimes be managed locally, at other times nationally or even trans-nationally. The challenge for "water governance" lies in achieving a balance between the principles of subsidiarity and global governance. Water must not be traded as a global good on an open market, because this would seriously undermine the right to water. Efforts are necessary to ensure access to water for all people and do guarantee an appropriate supply, embedded in local realities, and combined with global actions and concerns (ibid., p. 570).

However, democratic and sustainable water governance does not only need the access to water as a human right but also the sustainable social intercourse with water as a human duty. Both are included in the nine Principles of Water Democracy from Vandana Shiva (2002, p. 35 and 36):

1. Water is nature's gift. People receive it freely from nature and owe it to her to use this gift in accordance with their sustenance needs, to keep it clean and in adequate quantity.
2. Water is essential to life and source of all species that have a right to their share of water on the planet.
3. Life is interconnected through water which connects all beings and parts of the planet through the water cycle. Therefore, we have a duty to ensure that their actions do not cause harm to other species and other people.
4. Water must be free for sustenance needs. We receive it free of cost and buying and selling it for profit violates the inherent right to nature's gift and denies the poor of their human right.
5. Water is limited and can be exhausted if used ecologically and socially unsustainably by extracting more water from ecosystems than nature can recharge and consuming more than one's legitimate share, given the rights of others to a fair share.
6. Water must be conserved and everyone has the duty to use water within ecological and just limits.
7. Water is a commons and is no human invention. It cannot be bound and has no boundaries. It cannot be owned as private property and sold as a commodity.
8. No one holds a right to destroy, to overuse, abuse, waste or pollute water systems. Tradable pollution permits violate the principle of sustainable and just use.
9. Water cannot be substituted and is therefore intrinsically different from other resources and products. It cannot be treated as a commodity.

11.5 Conclusion

The sustainable treatment, beyond the Baconian dominance over nature, would have to leave, as a first step, the level of abstraction of a homogenous global natural good to be handled in the same way all over the world. Abstract references to "water" cover a number of aspects. Whereas there is "open access" to oceans, rivers are usually subject to national governments. Supplies of clean water and sewerage systems, on the other hand, are usually organized at the communal level. At the same time, water does not meet the same human needs everywhere. The goals of someone who relies on water to sustain life are different from the goals of those who want to make a profit from supplying water and sewerage services. The water itself cannot be subjected as some abstract generality to a technological apparatus for dominating nature, but must be perceived, accepted and treated in its specific qualities and manifestations.

And the sustainable treatment beyond the Lockeian programme of private property? Well, today, the final purpose of the global civil society is the preservation of natural commons and to achieve sustainability. If national governments through local and global governance is not able to protect natural commons and to achieve sustainability, it can be deposed.

References

Adler MJ (1996) Great books of the western world, vol. 28. Encyclopædia Britannica, Inc., Chicago
Bacon F (1990) Neues Organon. Lateinisch-deutsche Ausgabe. In: Krohn W (ed and introd). Felix Meiner, Hamburg [1632]
Dams and Development (2000) Report of the World Commission on dams. Earthscan Publications, London
Enquete-Kommission (2002) Globalisierung der Weltwirtschaft. Herausforderungen und Antworten. Drucksache, Berlin
Feeny D, Berkes F, McCay BJ, Acheson JM (1990) The tragedy of the commons: twenty-two years later. Hum Ecol 18:S1–S19
Goldman CR, Evoy JM, Richerson PJ (eds) (1973) Environmental quality and water development. W.H. Freeman, San Francisco, p. 80
Hardin G (1968) The tragedy of the commons. Science 162:1243–1248
Hardin G (1978) Political requirements for preserving our common heritage. In: Brokaw H P (ed) Wildlife in America. Council on Environmental Quality. Washington, DC, pp. 310–317
Kaul I, Conceicao P, Le Goulven K, Mendoza RU (2003) Providing global public goods: Managing globalization. Oxford University press, Oxford
Kaul I, Grunberg I, Stern MA (ed) (1999) Global public goods. International cooperation in the 21st century. Oxford University Press, New York/Oxford
Locke J (1823) Two treaties of government. Cambridge University Press, London [1690]
Omosa EK (2005) See Footnote 7: August 2005, International Institute for Sustainable Development (iisd). Pdf: www.iisd.org/pdf/2005/security_pastoral_water_impacts.pdf
Ostrom E, Dietz T, Dolsak N (ed) (2002) The drama of the commons. Natl Academy Pr, Washington, D.C.
Pateman C (1988) The sexual contract. Stanford University Press, Stanford
Petrella R (2000) Wasser für alle – ein globales Manifest. Rotpunktverlag
Scherhorn G (1998a) Privates and commons – Schonung der Umwelt als kollektive Aktion. In: Held M, Nutzinger HG (eds) Eigentumsrechte verpflichten. Individuum, Gesellschaft und die Institution Eigentum. Campus, Frankfurt am Main, pp. 184–208
Scherhorn G (1998b) Der Mythos des Privateigentums und die Wiederkehr der Commons. In: Biesecker A, Elsner W, Grenzdörffer K (eds) Ökonomie der Betroffenen und Mitwirkenden. Centaurus, Pfaffenweiler, pp. 29–42
Shiva V (2002) Water wars. Privatization, pollution and profit. Pluto Press, London

Study Commission on Globalization of the World Economy – Challenges and Responses (2002) Summary of the final report, June 2002. German Bundestag 14th legislative term

von Braunmühl C, von Winterfeld U (2005) Sustainable governance. Reclaiming the political sphere. Reflections on sustainability, globalisation and democracy. Wuppertal Institute, Wuppertal

Chapter 12
Towards a New Architecture of Agricultural Trade in the World Market

Wolfgang Sachs and Tilman Santarius

Abstract The reform of agricultural trade rules is at the center of negotiations at the World Trade Organization (WTO) regarding a multilateral framework for the global economy. The authors of this article, however, argue that the reforms envisaged do not bold well for the future of agriculture across the globe. According to them, they will deepen the desperation of farmers across the world and undermine local and global ecosystems. In contrast, this article explores new directions for trade rules beyond the free trade paradigm. Placing the challenges posed to agriculture and rural communities at the center of attention, it proposes political perspectives and policy instruments for a trading system that offers genuine opportunities for the poor, preserves the environment, and helps agriculture leap into the post-fossil age.

12.1 Enlarging National Policy Space

Trade liberalization empowers transnational business by disempowering national politics. It follows the philosophy that state failures by far surpass market failures when it comes to promoting the common good.[1] Consequently, structural adjustment programs as well as obligations under the World Trade Organization (WTO) and other free trade agreements have sought to restrict the scope of national politics through the regulation of trans-border flows in order to remove barriers to

---

W. Sachs (✉)
Wuppertal Institute, Wuppertal, Germany
e-mail: wolfgang.sachs@wupperinst.org

[1] The article presents modified parts of a document which was published by the organizations Misereor, Heinrich-Böll-Foundation, and The Wuppertal Institute for Climate, Environment and Energy: "Slow Trade – Sound Farming. A Multi-lateral Framework for Sustainable Markets in Agriculture." It was written by Wolfgang Sachs and Tilman Santarius in collaboration with Souleymane Bassoum, Daniel De La Torre Ugarte, Gonzalo Fanjul Suárez, Anna Pijnapple, Arze Glipo, Aileen Kwa, Hannes Lorenzen, Sophia Murphy, Odour Ongwen, and Rita Rindermann and printed in Germany in April 2007 (96p). It can be downloaded from www.ecofair-trade.org

the free flow of goods and investments. Moreover, as the WTO considers barriers not only tariffs or quotas at the border, but nontariff measures as well, such as price controls, investment rules or health standards, the power of societies to protect the public interest according to their collective preferences is seriously weakened (Wade 2005). This contradicts the principle of Democratic Sovereignty that recognizes a society's right to self-governance and diversity. More specifically, when faced with the pressures of trade deregulation, governments tend to downplay the importance of providing for universal access to social and environmental common goods.

Indeed, it is hard to see how essential public benefits can be effectively provided for unless politics assumes responsibility for it on the national and sub-national level. For instance, safeguarding the human right to food may call for reviewing land tenure laws. On the other hand, redirecting farming practices towards a regenerative agriculture may require a particular system of economic incentives and disincentives, or linking crop cultivation and industry may imply changes in investment policies. In particular, with regard to livelihood rights, environmental protection, and sustainable economic development, it is only in the national space where situation-specific policies can be implemented, which not only grow out of the political consensus, but which are grounded in local knowledge and commitment. Without a certain amount of ownership by the political community, it is likely that common goods will not be adequately protected, thereby twisting the welfare balance of trade into the negative. Similar arguments underlie the 2004 UNCTAD XI concept of "political space," which, however, tends to refer only to developing countries (South Centre 2006).

From this perspective, in particular for developing countries, it is clear that trade reforms are misconceived if they privilege export promotion over import governance; the management of imports is more important for the well-being of a society than the facilitation of exports. The question is not – as in the wake of the Washington Consensus – which countries need to be integrated into the world market, but what they need to achieve equitable and sustainable development. Since less powerful economies are particularly vulnerable to cheap and unqualified imports, they must be guaranteed the right to regulate access to their internal markets in order to best protect their human development needs. It was always a mistake to believe that fairness in world trade is achieved simply by providing better access for the South to Northern agricultural markets. Instead, what matters more is the ability of weaker countries to regulate imports in order to protect, if necessary, young industries, small farmers or indeed a fragile environmental base.

However, it has to be admitted that the plea for more national policy space can become counterproductive in the context of authoritarian or corrupt governments. Unfortunately, there are a large number of governments that are not governed by democratic regimes or where they formally are, they continue to promote policies that serve elite minorities instead of the majorities of citizens. And in other cases, many governments simply do not function efficiently and lack the institutional capacities to implement effective policies. However, policy space is the basis for domestic social forces to demand and secure their democratic rights.

12.2 For the Sake of Livelihood Security

As agriculture remains the main source of livelihood for the majority of populations in most developing countries, the sensible policy for governments and other policy makers is to ensure that import liberalization takes a back seat when domestic livelihoods and food security are at stake. More so, in the light of the human right to food, political authorities have the obligation – and consequently the right with respect to international rules – to protect, sustain and support the necessary conditions to encourage production of sufficient healthy food in a way that conserves the land, water and ecological integrity of a place, and respects and supports producers' livelihoods (ICFFA 2003).

Above all, this obligation requires adequate space for the governance of imports to protect small farmers and artisans from devastating import surges. For many developing countries that have lost this space as a result of structural adjustment, they are often unable to control the volume of cheap food or dumped products that are flooded into their markets. But the availability of cheap imports will not ensure national food security if domestic agricultural production is undermined by imports of food or nonfood crops. For example, in India accelerated imports of edible oil products drove away countless producers of sunflower, coconut and palm oil; in Ghana stockbreeders and butchers cannot survive against the massive volume of cheap meat imports from Europe; and in Mexico maize farmers have been driven to the wall by subsidized exports from the United States. Imports that undercut domestic prices reduce consumption costs for urban dwellers but undermine the livelihood of countless people engaged in agriculture and food production. In such circumstances, it is better for governments to restrict trade, rather than to put a large number of rural livelihoods at risk (Malhotra 2006).

While there have been attempts by governments to raise border protection, the remedies available within the WTO and bilateral agreements are limited. The various agreements in the WTO, including the GATT and the Agreement on Agriculture, theoretically offer Member States some options for safeguard measures and quality standards. But in practice, these have been of limited use to countries and have proved to be completely inadequate in addressing price volatilities in the international market (Glipo 2006). In the Doha Round of negotiations, countries of the "Group of 33" introduced a Special Safeguard Mechanism that was both price and volume-triggered; however no proposal of quantitative restrictions on imports has been placed on the agenda. Indeed, the suggestion for a safeguard mechanism was undermined by the central focus of the Round on unrestrained market access in both North and South. Such a focus implies that tariffs or other instruments of border protection are gradually removed rather than redefined in order to create a space for the domestic economy to develop. In contrast to open border politics, this report maintains that countries should provide more policy space to use both tariffs and quantitative restrictions. This may include price- and volume-triggered tariffs, price bands, as well as quantitative restrictions (e.g., quotas), or other safeguard mechanisms (Glipo 2006, Malhotra 2006).

12.3 For the Sake of Sustainability

In addition to securing farmer livelihoods by protecting them from devastating import surges, countries need policy space to implement policies and measures that chart their self-defined path to sustainable development. This is consistent with the principle espoused in the Johannesburg Plan of Implementation of the World Summit on Sustainable Development that "each country has the primary responsibility for its own sustainable development, and the role of national policies and development strategies cannot be overemphasized" (WSSD 2002). Following the proposed principle of Economic Subsidiarity, such an approach calls for policies that make domestic production and processing of food a priority, along with the development of domestic markets. Furthermore, following the principle of Environmental Integrity, it calls for policies that discourage pollution and overuse of soils and water, while encouraging the transition towards a biodiversity-based agriculture. In the light of sustainable development objectives, the governance of imports is not just a matter of restricting cheap imports, but rather of linking the import of goods, services, and capital to sustainability considerations. Countries must retain authority, for instance, to influence flows of foreign investment, to direct the activities of transnational corporations, to link domestic production to strict social or environmental standards, or to design support schemes to ensure healthy rural economies.

National regulation has been increasingly impeded in the past two decades by the introduction of structural adjustment policies in the 1980s, by the increasing number of side agreements to the GATT, and later on by the WTO, including the Agreement to Technical Barriers to Trade (TBT), the Agreement on Sanitary and Phytosanitary Measures (SPS), the General Agreement on Trade in Services (GATS), or the Agreement on Trade-Related Investment Measures (TRIMS). As these and other agreements restrict technical regulations, domestic support measures, or the implementation of health and social standards, public policy loses its capacity to support society in protecting public goods. Furthermore, as the GATS – and as well, some bilateral and regional agreements, such as the North American Free Trade Agreement (NAFTA) – extend the nondiscrimination principle beyond products and services to actual companies, this creates serious problems precisely in the regulation of (agricultural) services, such as banking or extension, which are central to livelihood security.

However, countries must assert their authority in restricting the activities of foreign corporations if they conflict with national anti-trust legislation, or if they abuse their market power for price manipulation or the building of cartels. Further on, national governments might need to regulate corporate activities in order to protect the interests of domestic producers. For example, contract-farming by foreign supermarkets could be conditioned to retain a fair share of the profits for local farmers. Governments might want to improve the inter-linkages between farmers, local processors, small-scale retailers, and consumers in their rural areas, as well as between foreign corporations and local economies, so as to keep as much value creation as possible in the region and to protect against the flight of capital. Therefore, policy space for local content policies or for legislation on requirements for joint ventures with local firms must be retained.

Likewise, policy space must be preserved for specific support measures. For instance, domestic supply management schemes will not function properly unless they are linked to effective border-control measures that restrict imports of those products under the scheme. The same holds true for state trading enterprises or state-owned marketing boards. If such institutions should substantially support farmers in food distribution and marketing, guarantee minimum prices, and stabilize price levels through buffer stocks or storage, they will require corresponding domestic legislation that controls import prices and quantities, and mandates such management of trade flows at the national level.

In addition, countries must be able to defend their right to impose measures for the sake of food safety, food quality, and environmental security, since these are important measures for preventing food-borne illnesses, and to protect the natural resource base and the resilience of ecosystems. This requires increasing the capacity of countries in developing not only stronger and more effective regulatory measures, such as process and production standards for sustainable farming, processing, and retailing, but also standards for the installation of monitoring and risk-assessment systems. It would defy the principle of Democratic Sovereignty if such domestic measures would become subject to a multi-lateral review mechanism that decides upon their necessity, let alone their legitimacy. Moreover, once countries have strict domestic legislation in place, they need adequate policy space to impose these same standards on imports. Countries must be empowered to condition access to their markets with certain sustainability considerations in order to prevent domestic producers from being disadvantaged by importers (Fig. 12.1).

Greater national policy space is needed:

- To protect small farming systems from import surges through border control policies, including tariffs, quotas, and price- and volume-triggered safeguard measures;

- To ensure the functioning of support policies, such as supply management or state trading enterprises, through selected border control measures;

- To allow for domestic regulation on food safety, food quality, and environmental security;

- To maintain a level-playing field between responsible domestic producers and importers through corresponding quality conditions on imports; and

- To implement guidelines for foreign corporations, including local content policies or conditions on foreign direct investment for increasing domestic value creation.

Fig. 12.1 Greater national policy space is needed

12.4 Investing in Multi-Functionality

Agriculture in most countries is in a fatal bind. On the one hand, farmers are struggling with declining farm income and corporate concentration, but on the other hand, they are expected to provide indispensable public benefits without remuneration. In particular, small and medium-scale farmers in fragmented ecological settings are highly vulnerable to competition pressure due to biased agricultural policies and subsidies towards industrial agriculture and malfunctioning market mechanisms in the global market. For this reason governments across the world are constrained to provide institutional or financial support to agriculture for securing food production and sustaining family farms. Except in countries with vast expanses of agricultural land and little traditional farming, small-scale and family farming is unlikely to survive unless it is supported by public policy measures. Furthermore, just as support is required to ensure the viability of social common goods, it is also required to underpin the provision of environmental common goods. Under competitive conditions, farmers must be rewarded for producing – as economists say – positive externalities, such as clean water, biodiversity, and rural landscapes. In both cases, it is the so-called multi-functionality of agriculture, which is at stake, and which distinguishes it from other business sectors.

Against this backdrop, the long-running debate on the reduction of domestic support for agriculture, the second pillar of reform under the Agreement for Agriculture, manifests itself in a new light. While a number of economists and politicians view domestic support policies as a stronghold of protectionism to be entirely dismantled, appreciation for the nonmarketable dimensions of agriculture suggests a shift in perception. For ensuring the multi-functionality of agriculture – in both the social and the environmental sense – calls for domestic support. Insofar as this insight is taken seriously, the search for fair and environmental trade rules changes in focus. Attention will be directed towards the proper level and structure of domestic support rather than its elimination.

Nevertheless, it goes without saying that the present systems of domestic support are woefully inadequate in promoting multi-functionality. As to its social side, agricultural subsidies in the US and the EU (both among the largest trading powers each of whom grant the highest levels of subsidies) flow mainly to large industrial landholders, retailers and food industry instead of to family farms and to sustainable rural development. Eligibility for subsidies in the US is not linked to income levels, but to the type of crops that farmers produce. Ninety percent of payments are channeled towards corn, wheat, soy, and rice, while farmers who produce about 400 other crops receive no financial assistance whatsoever (Baldwin 2005). In the EU, since the latest reform of the Common Agriculture Policy (CAP), direct payments to farmers on a hectare basis allow companies to include these subsidies into their price calculations, for example for machinery and chemical inputs but also for low farm gate prices the processing industries pay. As the hectare-based payments are in most cases not bound to employment or environmental conditions, 80% of the total subsidies continue to be accumulated by less than 20% of the farms (Baldwin 2005). This is why intensive, large-scale farms and export-oriented agribusiness profit most from the public payments. As to the environmental side of multi-functionality, a similar story holds true. As public funds continue to be used to intensify agricultural production the effect is largely the decline of ecosystems. Subsidizing chemical inputs, machinery, irrigation, and factory farms externalize negative effects on the environment and costs on society as a whole.

Against this background it is high time to redesign current domestic support schemes. A first step in this endeavor is to clearly distinguish between at least three different types of support. The first type is market price support in which producer and consumer prices are influenced through a range of policies, such as guaranteed prices for certain produce, tariffs and levies on imports, or quotas, among others. Market price support has not only come under criticism because such "dirigiste" measures are not compatible with the free trade paradigm; their essential shortcoming is that they provide incentives to over-produce, and hence contribute to dumping and a depression of prices in foreign markets. Introducing supply management schemes where possible is a viable solution: they stabilize prices, but without creating over-supply.

A second type of domestic support consists of direct payments to farmers in which money is transferred from taxpayers to producers without raising consumer prices. In the past, due to WTO requirements both supply management schemes as well as other market price support measures have been largely replaced by increasing amounts of direct payments. Yet such payments create problems of their own. As the increasingly concentrated agro-industry can indirectly use these subsidies to lower farm gate prices, this still provides incentives for increased production, as farmers may continue to produce even if they are not competitive. Therefore, direct payments must strictly be conditioned to improve sustainable production practices, create employment, and reduce dumping practices.

A third category of domestic support consists of specific support measures for rural economies, such as research, extension, education, infrastructure, as well as rural development and agro-environmental programs. Oriented in the right direction, so as to foster environmentally benign small-scale and family farming, this category of support can combine policies and measures that create an "enabling policy environment" for sustainable agriculture (Pretty 1995). Considering that farmers should receive most of their income from their farming, and not from the government, a combination of ecologically and socially conditioned direct payments, supply management as well as an enabling policy framework should guide the reform of domestic support schemes.

12.5 Policy Frameworks for Sustainable Family Farming

The guiding strategy for governments intending to facilitate access by family farmers to internal markets – which matters more than their access to foreign markets – should be to support small farmers, in particular women, in reclaiming long-term access to their domestic and local markets. First and foremost, this includes policies beyond trade, which protect the land rights of communities as well as access to basic natural resources, especially strengthening women's rights and land entitlements. Moreover, as countries step back from both export-orientation and import dependency, governments will need to ensure that decentralized rural infrastructure fosters local marketing. They must also ensure that rural and urban areas are sufficiently connected so as to elevate the hinterlands as the main suppliers of food for towns and cities.

Additionally, if small-scale production should be favored over large-scale monoculture farming, these farmers will require support achieving "critical economic mass" through associative forms of economic activity, covering, for instance, joint warehousing, processing, and marketing. A good example is the Anand Milk Producers Union (AMPU) in India, which was so successful that the National Dairy Development Board of India adopted it as its model. Owned by a union of small milk producer cooperatives, which are in turn owned by hundreds of rural women – some of which actually own one dairy cow, the Union operated a large, modern dairy facility that supplied a variety of quality branded dairy products all over India (Ongwen and Wright 2007, Korten 1999). Governments should provide institutional and financial support, including public finances for micro-credit and loan programs, to foster such associations.

In this vein, governments are well-advised to empower farmer organizations and producer cooperatives to help them play a decisive role in the local and regional market. In several parts of Latin America, for instance, the direct participation of family farmers in the local market has been improved through self-initiative and NGO-support for the creation of weekly ecological markets (Ferias Ecológicas). Relatively small infrastructural and knowledge support – for example, provision of market stands or timely transport, support with advertisement and training in basic bookkeeping – have had tremendous impacts. Similarly, several successful initiatives of local and regional trade networks have emerged, also in industrial countries.

In addition, improvements in small-scale production depend far more on expanding the knowledge base than on expanding the amount of farm inputs. Indeed, analysis has shown that in those countries with successful increases of agricultural productivity, public investments in agricultural research and development as well as in rural infrastructure were the most important drivers.

Besides calling for a reorientation of research and development, governments, research institutions as well as farmer cooperatives should advance low-cost, locally specific technological development that improves both the productivity and the environmental sustainability of more extensive and traditional knowledge-based farming systems. Research should be re-oriented towards the needs of small-scale and family farmers and sustainable agriculture, and it should become more farmer-led. In addition, research should professionalize the exchange of traditional knowledge, in particular for female farmers, because in times of global environmental change and fast-evolving economic restructuring, traditional knowledge on seed breeding, sustainable farming practices, and small-scale marketing strategies must be constantly improved by inter-cultural learning and information sharing.

Finally, farmers should be supported in their constant transition towards more sustainable farming practices. In North and South alike, farmers will need to maintain their natural production base and to produce healthy quality products in order to remain viable in the long-term. Multiple strategies to de-industrialize agriculture have been developed during the past decades, including Resource-Conserving Agriculture, Organic Agriculture, as well as Agroecology as the most effective way of restoring on-farm nutrient cycles and establishing biodiversity farming practices (Sachs and Santarius 2007).

Governments must support this transition through a range of policies and measures that have proven viable in the past (Pretty 1995). For instance, if polluting practices are penalized with taxes and levies, this makes polluters pay for the resulting environmental costs, and hence, will reduce pollution. Taxes could also be raised on industrial farm inputs, such as fertilizer or pesticides, so as to accelerate the transition towards closing on-farm nutrient cycles. If farmers' training and farmer field schools for sustainable farming practices are supported, and if the capacities of respective local NGOs are scaled up, this will catalyze further activities in the farming communities and generate local ownership in the process. Most importantly, however, governments should foster the development of local and civil society-based schemes for sustainability process and production standards, and develop strategies to ensure that standards are mainstreamed in all aspects of agricultural production.

12.6 Tight Conditions on Direct Payments

As supply management schemes and enabling policy frameworks correct the market trends currently working against sustainable family farming, they make compensatory payments to farmers much less necessary. Nevertheless, limited governmental subsidies might be needed in certain cases. For instance, as farmers face real adjustment costs when converting to a more sustainable agriculture, governments may need to provide subsidies for the transition period. However, the current schemes of massive direct payments must be reformed, and any type of direct payments should be conditioned to strict criteria. Since currently most payments still maintain an incentive for the maximization of yields, and thus for over-production, they must be profoundly reformed. In addition, eligibility for direct payments should be made dependant on the application of sustainable farming practices, while the amount paid should be linked to the number of jobs offered on the farm (Reichert 2006). This will promote rural employment and benefit farms that carry out labor intensive and environmentally sound farming practices.

12.7 Support Without Dumping

In the current debate, governmental support schemes are usually blamed for two reasons: first, support is said to distort prices and increase domestic production, and thereby decrease the market share of imports. Secondly, support is said to cause product dumping onto other markets. In the context of an eco-fair trade regime, and for the sake of the principles of Democratic Sovereignty and Economic Subsidiarity, the former concern is not a priority. For no society in the world, be it in the South or the North, should be prevented from achieving food self-sufficiency on their own terms. However, a multi-lateral trade regime that respects the principle of Extraterritorial Responsibility should ensure that support schemes do not harm others. For the dumping of products, either through export subsidies, in the worst case scenario or through green box payments, is in any case illegitimate. As a stop-gap measure until agricultural dumping is effectively prohibited, a multi-lateral institution should be authorized to establish a "Dumping Alert Mechanism" that warns governments when dumping may undermine farmers' affairs in the importing countries. On the basis of that information, importing countries should be advised and provided the opportunity to protect their domestic sector, for example by adding a percentage tariff equivalent to the dumping margin on their tariff levels (Fig. 12.2).

Elements of a 'Dumping Alert Mechanism':

- A Dumping Alert Mechanism warns governments when dumped exports may undercut farmers' livelihoods in importing countries;

- Exporting countries are registered at a multilateral body and are required to provide information on each year's support levels;

- The multilateral body verifies this data and publishes for each exporting country the amount of dumping that takes place;

- Countries that import goods are informed and advised to increase their border tariffs vis-à-vis countries that practice dumping.

Fig. 12.2 Elements of a "dumping alert mechanism"

However, the world market price might be too low to serve as a point of reference, especially when major suppliers fail to incorporate the costs of social and environmental damage. For instance, neither the costs of the irreversible depletion of ground water for irrigation from fossil aquifers in the US Midwest, nor those arising from the deforestation of primary forests for pasture land and, successively, export-oriented soy monocultures in the Brazilian Cerrado, count into the support calculations of the WTO, or the OECD. Apart from this fundamental flaw, it has to be acknowledged that the full costs of sustainable agricultural production – in a world of highly diverse social settings and ecosystems – can only be defined in a national (or even regional) context, not at the global level. What it may cost to sustain family farming and the natural resources base in a region with prime conditions may not be sufficient to sustain farming systems in a region with marginal land. Hence, the ideal of a global "single price" that maximizes efficiency across all economies is incompatible with the principles of sustainability.

Against this backdrop, it is important to consider the additional impacts of dumping products that are sold at artificially low prices where they do not internalize the full environmental and social costs of production. In an eco-fair trade regime, a product would be considered to be dumped if it was sold below the market price in producer countries that internalized social and environmental costs. This new concept of dumping would prevent current trends of mounting cost externalization. Even conventional economic theory predicates "free trade" on the basis of full production costs, which exclude social and environmental externalities.

12.8 Stabilizing Prices to Protect Farming Livelihoods

The predominant problem for agricultural producers around the world is the declining world market price for food staples. Family farmers everywhere, be they poor or prosperous, be they Southern or Northern, suffer from drastic price variations and all-time low prices that depress their income and threaten their livelihoods. Societies should in any case protect their farm sectors against import surges and they should promote support for sustainable family farming. However, these measures will not be enough to stabilize global price levels as long as other countries continue to oversupply the world market. Moreover, one of the main factors behind low farm gate prices is not over-production, but corporate power and control of the market. In what is called a buyers' market, powerful processing or trading companies can set prices at will, and hence continuously depress farm gate prices. International trade negotiations must address the problem of world price volatility and price decline as a matter of highest priority.

12.9 A Cooperative Mechanism for Balancing the World Market Supply

In the medium-run, given advances in crop yields and the increase in crop acreage in countries like Brazil due to the persistence of intensive and export-oriented agriculture, there will be a need for the major crop exporting countries of the world to establish cooperative mechanisms to manage the production of crops (Rosset 2006). At the multi-lateral level, negotiations could be launched to adopt a "Multi-lateral Cooperative Framework for Balancing the World Market Supply." This framework would leave the actual implementation of supply management schemes to domestic policy makers. The multi-lateral framework would not only ensure that major exporting nations implement supply management schemes, it could also solve the "prisoner's dilemma," namely that world market supply management can only be achieved cooperatively.

Currently, the world market in food staples, such as cereals and oilseeds, as well as for products like cotton, sugar, or rice, is dominated by merely a handful of countries. Therefore, a multi-lateral framework that includes the main exporting countries of these crops would be viable and enforceable. For example, six countries – Argentina, Australia, Brazil, Canada, the EU, and the US – held 47 and 58% shares respectively in global production capacity of wheat and corn, and 52 and 64% shares respectively of global exports in 2003 (FAOSTAT 2006). A multi-lateral framework with these countries as main parties would indeed be a considerable contribution to a fairer distribution of production capacities and, therefore, to poverty reduction and the economic renewal of rural economies worldwide (Fig. 12.3).

Steps towards a 'Multilateral Cooperative Framework for Balancing the World Market Supply:'

- Identify those countries with a significant influence on world market prices as participants of the scheme (e.g. Argentina, Australia, Brazil, Canada, EU, US etc.);

- Agree on the crop-specific caps that would govern overall global production capacity (e.g. -3% of global wheat production) in order to raise world market prices above a certain minimum level;

- Determine country and crop-specific reduction targets (e.g. US -8%, EU -4% etc.), according to each country's share in global exports;

- Implement monitoring and verification mechanisms to assist countries with compliance (e.g. independent third party verification); and

- Ensure flexible review of the scheme over short periods for adjustment and improvement, and for improving implementation at the national level.

Fig. 12.3 Steps towards a "multi-lateral cooperative framework for balancing the world market supply"

12.10 Setting Standards for Quality Trade

In agriculture and many other economic sectors, the present-day economic system is anything but a least-cost system (Hawken et al. 1999). In a true least-cost system, the losses inflicted upon common goods while producing commercial goods would be weighed against the gains made in the market. From this viewpoint, the objective of agriculture is not just to produce earnings but to contribute to the health for all, including nutrition for people and regeneration of natural ecosystems. Food systems, therefore, are to be evaluated in terms of a common health framework that accounts for both the quality of food and the long-term health of communities and ecosystems (Dahlberg 2002). However, since the free play of market forces favors private gain over common goods, it is up to politics to rectify this imbalance. Public policy interventions are necessary to ensure framework conditions that align the pursuit of private gain with the protection of the biosphere and human rights.

Moreover, trade reform has to create a level playing field in the social and environmental responsibility between farmers and businesses. At present, deregulation unduly favors unsustainable farming practices and trading decisions, since corporations locate activities where social and environmental costs can be most easily externalized. Too often, the dismantling of protectionism has resulted in protection of the ruthless. For instance, sugar workers in Brazil toil while supermarket chains compete at low prices, moreover, the elimination of mangroves may optimize shrimp production for middle class dishes while creating environmental hazards, and finally, pesticides used in Pakistani cotton fields, while polluting soils and laborers is indeed the hidden price for easy shopping in the fashion stores of the world. As long as production costs are not required to incorporate the cost of safeguarding common goods, free trade will continue to accelerate both the marginalization of the poor and the decline of the biosphere. It is only through minimum standards for securing the dignity of labor and the integrity of the global environment that a groundwork for a fairer and safer twenty-first century can be established. In the end, trading internationally must be understood as a privilege to be offset by internalizing social and environmental costs.

12.11 Sustainability Process and Production Standards

As a first step, national politics should foster the development of standard monitoring and verification schemes. The establishment of production process standards is crucial for minimizing clear-cutting, over-exploitation of water reserves, chemical pollution, or greenhouse gas emissions. The feasibility of monitoring and evaluating production processes has been clearly demonstrated by fair trade and organic agriculture initiatives, which are usually enforced by inspection and certification bodies. The "IFOAM Norms" for organic agriculture, as one example, include a detailed set of general principles and standards with requirements for crop production and animal husbandry, including criteria for the evaluation and use of selected off-farm inputs, and standards for processing, handling, and labeling (IFOAM 2002). Although IFOAM is considered as the global platform of the certified organic movement, the IFOAM Norms are but one set of standards among many others that have been developed by national or private organizations. Today in more than one hundred countries, farmers' organizations and consumer groups have developed their own sets of organic standards and certification rules – many of them consistent with IFOAM provisions, but specified and adapted to their respective environmental and social circumstances (Barrett et al. 2001). Governments should support the independent development of such standard schemes.

In a second step, governments should plan to develop domestic agricultural transformation strategies with standards becoming mandatory for all agricultural production. The steep increase in the volume of global area farmed under certified organic agriculture (IFOAM 2006) has resulted in significant environmental and social improvements. For instance, organic agriculture consumes less water and generates less soil pollution and fewer health risks. At the same time, species diversity is on average 30% higher than in conventional farming systems. In many cases, it is more labor intensive, because more sustainable soil management crop rotation, associated crops, sustainable weeding practices and precautionary pest treatment practices substitute chemical pesticides through labor (Maynard and Green 2006, Dabbert et al. 2002, Offermann and Nieberg 2000).

And yet critics argue that environmental standard schemes for production processes are socially imbalanced. For certification can be costly and complicated and, therefore, tends to disadvantage small producers. Costs can be reduced if farmers form producer groups or co-operatives that are certified as a whole; but fees may still be high, and internal inspection systems within the group create new costs. Therefore, given the fact that quality control is necessary, governments should foster the development of local, independent sustainability certification schemes. Local schemes have the potential to establish monitoring and certification mechanisms that are best suited to the structure of the farming system and the economic capabilities of the farmers; they can best minimize costs and regulatory burdens placed on small producers.

Moreover, locally and nationally independent schemes could be supported by a mechanism that shifts the costs of certification from farmers engaging in sustainable production to those who continue conventional practices, as well as from farmers to consumers. The experience with energy feed-in laws, which catalyzed an impressive penetration of costly renewable energy systems in the energy market in several countries are models that could be considered in the agriculture context. For example, a fee could be added on all conventional products, which in turn cross finances the costs for certification in sustainable agriculture, and assists small farmers in complying with standards and certification requirements.

12.12 Qualified Market Access

A trade regime that is serious about sustainability should support such inclusive sustainability standards at the national and international level. On the basis of proven implementation of domestic sustainability process and production standards, governments must have the competence to also link market access to these standards. Thus, trade in more environmentally and socially sound products will be favored over trade in conventionally produced goods. Indeed, the qualification of market access in terms of social and environmental requirements is urgent since agro-industries and food retailers increasingly invest in countries where environmental and social requirements are weakest. Such a strategy transforms these actors into protagonists of unconditional market access in countries with high food prices, thus increasing profits from sales, but undermining the competitive position of domestic responsible producers. Sustainability standards at the border would work like trade filters to reduce social and environmental dumping (Lorenzen 2007). Governments could provide a "carrot" to sustainable producers and grant preferential market access to products that adhere to certain sustainability standards (Clay 2004). In other words, commercial goods that have been demonstrably co-produced along with common environmental and social goods would be given a trading advantage, thus encouraging a shift in production and marketing towards eco-fair commodities worldwide.

First, it is probably a misconception to believe that Northern countries will be less offended than Southern countries by standards that aim at de-industrializing agriculture. This might be the case today since standards mostly encompass elaborate hygiene or health requirements for products. However, it might be different when Northern as well as Southern countries develop interests to protect their markets from social and environmental dumping. Any move towards sustainable agriculture will be doomed to failure if cheap foreign goods produced by destructive methods are allowed to penetrate the market. In this regard, it will also be up to the North to change its practices. It is not inconceivable that one day India will produce its own environmental production standards for poultry imports or Thailand standards for sustainable fishing. To be sure, the spread of industrial agriculture is global in scope, and even in poor countries, regions that are well integrated into the global market are usually characterized by industrial agricultural production systems. However, overall agriculture in the North is much more industrialized than in most of the South. For instance, the level of mechanization is nearly four times higher in developed countries than in developing countries. Regarding the use of synthetic fertilizers – and presumably pesticides, too – this picture is less clear. Still apart from China, Brazil, India and a few other developing countries, the majority of the developing world uses less fertilizer than the developed countries (FAO 2005). Moreover, many countries of the South with their vast regions characterized by small-scale agriculture that is organic by default will be better positioned than the countries characterized by industrial monoculture farming in most of the North.

It is neither regions nor farming systems, but only exports produced by environmentally harmful farming practices, which would be challenged through qualified market access. Therefore, a key question that must be addressed is where do such exports originate and who profits from low standards? Although extensive data is still lacking on the issue, presumably the bulk of global exports originate from high-input industrial systems in the North as well as from a few regions of the South (Sachs and Santarius 2007). For instance, the top five wheat exporters are the US, France, Canada, Australia, and Argentina – countries that are characterized by highly industrialized agricultural systems. If all EU wheat exports are included, about 75% of world exports in wheat from 2006 through 2015 will be produced by high-input farming (Vocke et al. 2005). Likewise the top three soy bean producers are the US, Brazil, and Argentina, which account for 80% of global soybean and 70% of global soy oil production (Ash et al. 2006). If their exports are challenged by the rest of the world through qualified market access, it will not be the small soy farmers in Brazil or Argentina who are affected, but the large industrial producers that account for the majority of exports from these countries. These producers as well, along with the respective transnational trading and processing corporations, have to be urged to shift towards more sustainable farming practices?

Furthermore, it has to be taken into account that in practice, more than governmental standards, corporate-created standards, such as EurepGAP, may become an unqualified trade barrier. Caribbean States have recently complained to the WTO against this initiative of European retailers which increasingly discriminates imports from developing countries based on food processing and long shelf-life standards. To the contrary, measures and instruments for qualified market access against ecological and social dumping would have to be developed simultaneously from the bottom up by civil society initiatives and from the top down by national governments. For instance, farmer networks like RIAF in the MERCOSUR region have initiated a mutual recognition of products from small farmers which are partly recognized by MERCOSUR member states as qualified for lower or zero tariff products; farmers mutually acknowledge themselves as "small farmers," while they are then recognized by other MERCUSOR member states that grant preferential access to products from these small farmers.

Finally, there is no question that the concept of qualified market access extends well beyond agricultural goods. The requirement that investments, goods, and services, which cross borders will have to measure up to social and environmental standards is an indispensable element for any eco-fair trade regime. The agricultural sector itself comprises a much broader range of goods than products, which are simply derived from plants or animals; the companies producing fertilizers, pesticides, machinery should be taken into account along with food processors and retail corporations. As well, the trans-border business of these companies must be subject to qualified market access. Why should Kenya not formulate investment standards for the entry of supermarket chains, Uruguay fuel standards for harvesters, or Thailand develop production standards for fertilizers? There is no doubt that qualified market access is neither to be restricted to agricultural goods nor to the South–North flow of trade. Quite to the contrary, given the unsustainability of developed economies, it is potentially more relevant for nonagricultural goods and the North–South flow of trade (Fig. 12.4).

12.13 Conclusion: Towards a Post-WTO Architecture of Agricultural Trade

GATT and WTO have been established on the basis of the principles of "Most Favored Nation" and "National Treatment," both expressions of the general principle of Nondiscrimination. In our view, nondiscrimination should continue to be an underlying principle so long as it is properly qualified by the principle of Democratic Sovereignty. Yet we suggest eliminating the National Treatment rule, at least in the context of agriculture. We believe that the ethos of global solidarity and the principle of Extra-territorial Responsibility require that nations should not be discriminated against, either in a positive or negative sense. However, they

> **'Qualified Market Access' and 'Sustainable Rural Development Fund'**
>
> - As a first step, countries would establish independent quality standards and certification systems at the domestic level. As a second step they would evolve these standards into mandatory requirements for domestic producers
>
> - Based on proven implementation of these mandatory requirements, countries could then gradually impose quality standards at the border and differentiate market access conditions between products that adhere to their sustainability standards, as opposed to products that are unsustainably produced;
>
> - Revenues from tariffs applied to harmful products in the North are channeled into an international 'Sustainable Rural Development Fund', which supports the transition towards sustainable farming practices and the implementation of qualified market access schemes in developing countries.

Fig. 12.4 "Qualified market access" and "sustainable rural development fund"

justify protection of domestic producers over foreign competitors at the border. In this light, we concur with the principle expressed in the 2004 Draft Peoples' Convention on Food Sovereignty that: "Food sovereignty becomes the right of people and communities to decide and implement their agricultural and food policies and strategies for sustainable production and distribution of food" (Windfuhr and Jonsen 2005). Indeed, the proposed policies to govern imports are grounded in the principles of Democratic Sovereignty and Economic Subsidiarity, which themselves are incompatible with the National Treatment principle.

Furthermore, the concept of "nontariff barriers" is difficult to reconcile with the principle of Democratic Sovereignty. The concept was introduced during the transition from the GATT to the WTO. It has led to major interventions in support policies, patenting rules, basic services, and property laws, extending the influence of trade rules into domestic politics far behind borders. But the weight given to the concept of "nontariff barriers" undermines the right of people and communities to organize their affairs – for example support to farmers, intellectual property rights, and land tenure laws – according to their preferences. The language of "nontariff barriers" has a reductionist effect; it boils down complex and diverse political arrangements to mere obstacles to trade. In accordance with the principle of Democratic Sovereignty, trade policy rule-making should not interfere with domestic politics, but should concentrate instead on market access issues and on quality standards for international exchanges.

However, the principle of Democratic Sovereignty is circumscribed by the right of other people and communities to their own right to sovereignty. In other words, the freedom of a nation ends where the freedom of another nation begins. This is where the principle of Extra-territorial Responsibility comes into play, i.e. nations have to be held accountable for the external trans-border effects of their policies that might harm other countries. The most obvious examples are export subsidies, domestic support influencing export prices, food aid etc. that lead to dumping on international and foreign markets. It is on the basis of Extra-territorial Responsibility that such policies must be abolished, and not on the grounds of establishing a global level playing field.

Moreover, the principle of Democratic Sovereignty is also circumscribed by the principle of Trade Justice. The latter principle, especially if understood as a systemic differential treatment of countries, seeks to address the drastic inequalities among nations in the world; it systematically privileges less powerful nations over more powerful ones and requires that rights and duties must be distributed unequally, i.e. according to respective needs and capacities.

Finally, any new multi-lateral institution on agricultural trade would have to be established under the auspices of the United Nations. Therefore, the foundational principles enshrined in the UN Charter would naturally govern the new trade institution. As a result, all of the UN instruments on human rights, most notably the UN Declaration on Human Rights would underpin the new trade institution as well. By contrast with the goal of economic efficiency that is currently the dominant objective of the WTO, the new multi-lateral trade institution would be governed by the principles of Human Rights, Environmental Integrity, Trade Justice, and Economic Subsidiarity. The goal of economic efficiency would step back to become one among other means available to maximize employment opportunities and to achieve decent livelihoods, as well as environmental security and social justice.

References

Ash M, Livezey J, Dohlman E (2006) Soybean backgrounder. Economic research service (ERS) of US Department of Agriculture, USA

Baldwin RE (2005) Who finances the Queen's CAP payments? CEPS Policy Brief, no. 88, December

Barrett HR, Browne AW, Harris PJC, Cadoret K (2001) Smallholder farmers and organic certification: Accessing the EU market from the developing world. Biol Agric Hortic 19: 183–199

Clay J (2004) World agriculture and the environment. A commodity-by-commodity guide to impacts and practices. Island Press, Washington, DC

Dabbert S, Häring A, Zanoli R (2002) Politik für den Öko-Landbau. Eugen Ulmer, Stuttgart

Dahlberg KA (2002) Green revolution. In: Munn T (ed) Encyclopedia of global environmental change, vol. 3. Wiley, Chichester, pp. 347–352

FAO (ed) (2005) Earth trends data tables: Agriculture and food. http://earthtrends.wri.org/pdf_library/data_tables/agr1_2005.pdf

FAOSTAT (2006) Statistical database of the UN Food and Agriculture Organization. Accessed 30 November 2006

Glipo A (2006) Achieving food and livelihood security in developing countries: The need for a stronger governance of imports. EcoFair Trade Dialogue Discussion Paper no. 2. www.ecofairtrade.org

Hawken P, Lovins A, Lovins H (1999) Natural capitalism. Creating the next industrial revolution. Little, Brown, and Company, Boston

IFOAM (International Federation of Organic Agriculture Movements) (2002) IFOAM norms for organic production and processing: IFOAM Basic Standards. IFOAM, Victoria

IFOAM (2006) The world of organic agriculture: More than 31 million hectares worldwide. Statistics and Emerging Trends 2006. IFOAM, Bonn

International Commission on the Future of Food and Agriculture (2003) Manifesto on the future of food. Florence, Region of Tuscany

Korten DC (1999) The post-corporate world: Life after capitalism. West Hartford, San Francisco, CA

Lorenzen H (2007) Qualified market access. How to include environmental and social conditions in trade agreements. EcoFair Trade Dialogue Discussion Paper no. 5. www.ecofair-trade.org

Malhotra K (2006) A sustainable human development approach to the role of exports in a national development strategy. EcoFair Trade Dialogue Discussion Paper no. 4. www.ecofair-trade.org

Maynard R, Green M (2006) Organic works. Providing more jobs through organic farming and local food supply. Study for the Soil Association. http://www.soilassociation.org/web/sa/saweb.nsf/ed0930aa86103d8380256aa70054918d/f194c3c4ae11f3578025716c00584962/$FILE/organic_works.pdf

Offermann F, Nieberg H (2000) Economic performance of organic farms in Europe. Organic farming in Europe. Economics and policy, vol. 5. University of Hohenheim (ed) Stuttgart-Hohenheim

Ongwen O, Wright S (2007) Small farmers and the future of sustainable agriculture. EcoFair Trade Discussion Paper no. 7. www.ecofair-trade.org

Pretty JN (1995) Regenerating agriculture: Policies and practice for sustainability and self-reliance. Earthscan, London

Reichert T (2006) A closer look at EU agricultural subsidies. Developing modification criteria. ABL and Germanwatch, Hamm/Berlin. www.germanwatch.org/tw/eu-agr05e.pdf

Rosset PM (2006) Food is different: Why the WTO should get out of agriculture. Zed Books, London

Sachs W, Santarius T (2007) World trade and the regeneration of agriculture. EcoFair Trade Dialogue Discussion Paper no. 9. www.ecofair-trade.org

South Centre (2006) Operationalizing the concept of policy space in the UNCTAD XI mid-term review context. South Centre, Geneva

Vocke G, Allen EW, Ali M (2005) Wheat backgrounder. Economic Research Service (ERS) of US Department of Agriculture, USA

Wade R (2005) What strategies are viable for developing countries today? The world trade organisation and the shrinking of "development space." In: Gallagher KP (ed) Putting development first: The importance of policy space in the WTO and IFIs. Zed Books, London, pp. 80–101

Windfuhr M, Jonsen J (2005) Food sovereignty – Towards democracy in localized food systems. ITDG/FIAN International, Chippenham, Wiltshire

WSSD (World Summit of Sustainable Development) (2002) Plan of implementation

Chapter 13
Epilogue: The Schweisfurth Foundation – A German Food-Ethics-Platform

Franz-Theo Gottwald

In 1985 Karl Ludwig Schweisfurth, a business man who had spent years actively working in the food sector, established the Schweisfurth Foundation. He had played an essential part in industrializing and globalizing the meat production sector in Europe. For decades he had been instrumental in turning the fulfillment of basic consumer needs into a way of maximizing profits through automation, specialization, and concentration.

Beginning in the 1980s, he became ever more aware of the fact that although meeting food requirements through industrial solutions – like those that meet other basic needs like clothing, secure housing, or a promising future for generations to come – function on a highly technological level, they do so in a rather uncivilized fashion. Two observations in particular strengthened his notion that the conventional, industrial manufacturing processes for goods and commodities meant to satisfy basic needs, had reached their apex worldwide. On the one hand, it appeared that in his field of work, technical and economic frameworks for animal husbandry had been optimized in such a way that no additional cost reduction was possible. Examination and analysis of industrial food production proved that depleted soils, polluted rivers and streams, and insufficient feed supplies for animals went hand in hand with immeasurable pain inflicted on the animals during transport and slaughter. On the other hand, it had become obvious as early as the middle of the 1980s that working conditions in all branches of conventional industry had become ever more devoid of meaning, which led to feelings of alienation, and workers calling in sick in ever increasing numbers.

These insights into the final stages of large-scale industrialization led to the establishment of the Schweisfurth Foundation as a Think-and-Do-Tank for a new ecological culture of agriculture.

F.-T. Gottwald (✉)
CEO Schweisfurth Foundation, Munich, Germany
e-mail: cthomas@schweisfurth.de

Since its inception, the Schweisfurth Foundation has organized and funded more than 1,300 projects in the fields of science as well as research and development of:

- Wholesome and natural foods;
- Organic and environmentally friendly methods of agriculture and natural and species-proper animal husbandry;
- Environmentally friendly energy supply;
- Environmentally friendly refuse disposal;
- Environmentally friendly building practices which produce livable housing;
- Working conditions tailored to humane standards.

The Foundation has continually supported continuing and adult education and training. Another aim has always been support of cultural endeavors, expressed in creative and gentle stewardship of present-day conditions and an awareness of future needs. The Schweisfurth Foundation has also regarded art as a way of raising consciousness for a creative and gentle way of dealing with the world at large.

Starting at the beginning of the twenty-first century, it became obvious that questions of agriculture and food ethics have taken on an ever-increasing urgency. The problems of worldwide hunger and scarcity of water are becoming more and more pressing issues, while dependency on oil in food production has been driving up production prices to dizzying heights.

Internationally and parallel to these economically and ecologically disastrous developments, the young discipline of Food Ethics has grown in an impressive way, which is best illustrated by the number of discussions about concepts and actual problems. Nevertheless, the Foundation is convinced that discussions will have to be fostered along the whole food chain: Obesity, traceability, agro-food biotechnology, dairy industry, transgenic plants, novel food, biofuels, world trade system, etc. For years NGOs have been lamenting politicians' lack of interest in the establishment of ethical tools that would alter these factors. There are reasons enough to make Food Ethics an important issue at universities and in schools in ethics, political science, economics and other faculties and departments.

This is the reason why the Schweisfurth Foundation decided to invest in organizing a Food Ethics Platform in Germany to bring together brilliant minds from the international food ethics discourse community with German scholars. Besides annual meetings, a series of publications is planned. The first one will be the book at hand, presenting international discussions and current information about general and concrete questions concerning food ethics. This book can be used as an interdisciplinary textbook for students and teachers and all those interested in raising consciousness for international and global perspectives of topics concerning food ethics.

In the coming years, the Schweisfurth Foundation will deal with the ethical dimension of technology applications in agriculture and food processing. In addition to biotechnology and nanotechnology, satellite-based precision agriculture will be one of its future topics.

Further, the food ethics platform will include the following:

- Contributions to the development and implementation of ethics management systems in food retailing.
- Setting standards for employee assessment of in-house company policies concerning ethics; these will primarily focus on large retail businesses.
- Development of communications rationales and future-oriented standards for customer assessment of external company policies in food retail businesses.

Additional plans include the following topics or tasks:

- Competitiveness and ethics in food retailing;
- Sustainable business practices;
- Business social compliance in food retailing;
- Ethics issues arising between international trade and public policies;
- Ethics issues of national and regional policies;
- Plant and animal ethics;
- Fair trade;
- Corporate social responsibility in food retailing;
- Ethical assessment of new technologies in agriculture and food processing;
- Stakeholder management in food retailing.

The platform serves as a basis for academic activities in these fields and areas of research, while also building bridges to activists in society and the economy who work toward the maintenance of corporate activities without endangering economic and social systems. It helps in the deliberation of values relevant to political and entrepreneurial responsibility with regard to food safety, the environment, and the rights of future generations. It investigates the requirements for societal responsibility as practiced by those food retail operations, which exceed standards set by law. Corporate Citizenship, Corporate Social Responsibility, and Corporate Sustainability as well as corporate contributions toward sustainable development in agriculture and food production are viewed from the paradigm of sustainable development ethics and transmitted to the appropriate panels made up of society stakeholders.

The Food Ethics platform collaborates with EurSafe (www.eursafe.org), the Center for Gastrosophy of the University of Paris-Lodron at Salzburg, Austria (www.uni-salzburg.at), and the Rhein-Neckar Forum of the Culinary Arts (www.kulinaristik.net) in Germany. It also works with the World Future Council (www.worldfuturecouncil.org) and the Global Marshall Plan Initiative (www.globalmarshallplan.org) to enhance the momentum of public attention concerning the urgency of food sovereignty and food justice.

In conclusion I would like to thank my co-editors Hans Werner Ingensiep and Marc Meinhardt for their longstanding and fruitful cooperation and Ms. Susan Safren, Senior Editor at Springer Publisher. I would also like to extend my gratitude to the Board-of-Trustees of the Schweisfurth Foundation for their unwavering support.

Index

Note: the letters 'f' and 't' following locators refer to figures and tables respectively.

A
"Adequate food," 4, 119–120, 124, 128, 130–132, 135
Agrarian romanticism, 11
Agreement on Agriculture (AoA), 121, 129, 132, 134, 187
Agricultural biotechnology, 52–53, 65–66
Agricultural trade and human right to food, 119–134
    adequate food access, CESCR view, 119–120
    case study
        case study in Ghana, *see* Ghana, rice trading/right to food
        case study in Honduras, natural/manmade disasters, *see* Honduras, rice trading/right to food
        case study in Indonesia, *see* Indonesia, rice trading/right to food
    food crisis
        reaction of the World Bank to, 133
        WTO Agreement on Agriculture, role, 134
    General Declaration of Human Rights/ICESCR, 119
    rice trade liberalization, a threat to producers
        biggest exporting countries, 120–121
        increased rice imports, impact/challenges, 121
        policy reasons to boost import surges, 121
        rice, main source of calories/income for peasants, 120
    right to food in times of globalization, 119–120
    States' obligations
        CESCR's statement, 120
        ETO, proposed by EED/FIAN, 120
        extraterritorial violations, 120
        strategies to realize right to food, 120
Agricultural trade in the world market, 185–202
    balancing world market supply, cooperative mechanism, 195–196
    direct payments, conditions on, 193
    "dumping alert mechanism," elements, 194f
    livelihood security, sake of, 187
    multi-functionality, investing in, 190–191
    multi-lateral cooperative framework, steps involved, 196f
    national policy space, need, 189f
    post-WTO architecture of agricultural trade, 200–202
    protect farming livelihoods, stabilizing prices, 195
    qualified market access/sustainable rural development fund, 198–200, 201f
    setting standards, quality trade, 196–197
    support without dumping, 193–195
    sustainability process/production standards, 197–198
    sustainability, sake of, 188–189
    sustainable family farming, frameworks, 191–193
Agrochemicals, 155
Agroecology, 193
Agro-food biotechnology, 67–83, 206
    EU regulatory texts/practices, precaution in EU regulation, 70–72
    and hazy compliance, 72–73
    precaution and sustainable development dimensions of a precautionary attitude, 77–83
    precaution lost its orientation, 74–75
    re-linkage, 75–77

Agro-food biotechnology (*cont.*)
  public debate on
    historical reconstruction, *see* Public debates on GM crops, historical reconstruction
    inadequate EU regulatory response, 70
AML, *see* Animal microencephalic lumps (AML)
AMPU, *see* Anand Milk Producers Union (AMPU)
Anand Milk Producers Union (AMPU), 192
Animal biotechnology, 105
Animal breeding, 105–106, 112
Animal dignity, 101–115
Animal ethics, 4, 104–109
  challenges
    animal biotechnology, 105
    battery farming, 105
Animal microencephalic lumps (AML), 106, 108
Anthropogenic climate change, 4–5, 137–139
AoA, *see* Agreement on Agriculture (AoA)
Applied ethics, 49
Argumentative *vs.* prescriptive versions of precautionary principle, 89–90
Arrozazo, 125–127
Aswan Dam, 177
Australian Meteorological Association, 139
Autonomy, 12–13, 18, 22–24, 69, 102

B
Balancing world market supply, cooperative mechanism, 195–196
Basic labor rights/standards, ILO, 144–145
Battery farming, 101–115
Bayh-Dole Act (1980), 68
Binding international standards, animal/plant health sectors
  CAC, 140–141
  IPPC, 141
  OIE, 141
  SPS agreement, aim, 142
  WTO's agreement on Sanitary and Phytosanitary Measures, aim prevention of misuse of standards, 141
Biodiversity-based agriculture, 188
"*Bioethics: An Introduction for the Biosciences*" (Mepham 2005), 2
Biofacts, 10
Biofuels, 1, 5, 153–163, 206
Biosafety, 71–73, 87–97
Biota, 18t, 24
"Biotech is playing God," 3

Biotechnology applications, 101–115
  core ideas
    AML, science fictions, 106
    animal breeding, oppositions by western countries, 105–106
    animal ethics, challenges, 105
    animal protection beyond pain and injury, 105
    violation of animal/human dignity, Hauskeller's example, 105
  integrity
    protecting and maintaining, 107
    respect for animal integrity, elements, 107
    Schmidt's interference, animal integrity, 108
    species-specificity, 108
    Swiss/Dutch discussion, 107
  integrity and dignity
    concept of inherent value, 108
    inherent value and proper nature, dimensions, 108
  intrinsic value and dignity
    instrumental value, 109
    Taylor's approach, 109
  need for new concept, 103–104
    animal ethics view, Manuel Schneider, 104
    dignité de la creature, Richter's view, 103
    "dignity of creatures" replaced with "integrity of living beings," 103
    genetic modifications, debate over, 104
    legislations for "dignity of creatures," 104
  species and *Eidos*, 110–111
  success of concept
    dignity, a "thick concept," 103
    "dignity of creatures," critics, 103
    'dignity' used by Declaration of Animal Rights (UNESCO), controversy, 101–102
    human dignity, debate on, 102
    "IgNobel Peace Prize 2008," citizens of Switzerland, 101
    Mitgeschöpflichkeit/integrity, less commonly in use, 102
    speech act theory, 102–103
    widespread use of dignity/Würde, reasons for success, 102
    Wolf, "manifesto" character, 102
  telos
    Rollin's concept, 111–112

"telos-approach" or
  "flourishing-approach," 112
treating animals adequately
  actions and attitudes, 114
  Balzer/Rippe/Schaber report, 114
  battery farming/biotechnology,
    aesthetics, 113
  dignity of animals, exclusions (EKAH),
    113
  genetic integrity, 113
Blueprints, 112
BMI, *see* Body-mass index (BMI)
Body-mass index (BMI), 20–21
Bovine Spongiform Encephalopathy (BSE),
  92, 141
Brot für die Welt (Bread for the World), 4, 120
BSE, *see* Bovine Spongiform Encephalopathy
  (BSE)
Businesses as social enterprises (Muhammed
  Yunus), 145

C
CAC, *see* Codex Alimentarius Commission
  (CAC)
Cancer, 20, 36t, 58, 114
CAP, *see* Common Agriculture Policy (CAP)
Cardiovascular disease, 20
CARNEL, *see* Cooperativa Agropecuaria
  Regional El Negrito Limitada
  (CARNEL)
Carrefour, 33
Cartagena Protocol, 88–89, 96
Case study, *see* Agricultural trade and human
  right to food
CESCR, *see* Committee on Economic, Social,
  and Cultural Rights (CESCR)
Childhood obesity reduction, strategies, 26
  "command and control" regulations, 26
  performance-based regulation, 26
City of the Sun (1623), Tommaso Campanella, 7
Civil disobedience, 148–151
Codex Alimentarius Commission (CAC),
  140–142
Coff, C., 2, 31–45
Command and control regulations, 26
Committee on Economic, Social, and Cultural
  Rights (CESCR), 119–120
Common Agriculture Policy (CAP), 190
Common-pool resources, *see* Commons
Commons, 5, 148, 169, 170–175, 180–181
Compensatory power, 26
Comstock, G., 3, 49–65
Condign power, 26

Conditioned power, 26
"Conservation Agriculture," 140, 143
Consumer-citizen, 32, 35, 37
Consumer concerns, 2, 31–32
Consumer-supported agriculture group
  (CSA), 2
Cooperativa Agropecuaria Regional El Negrito
  Limitada (CARNEL), 127
Co-responsibility, 82
Corn, 5, 146, 154, 157, 190, 196
Corporate social responsibility (CSR), 163, 207
CSA, *see* Consumer-supported agriculture
  group (CSA)
CSR, *see* Corporate social responsibility (CSR)

D
Deblonde, M., 3, 67–83
Declaration on basic principles and rights at
  work (1998), 144
Democratic Sovereignty, 186, 189, 194,
  200–202
  principles of, 194, 200
Department for Innovation, Universities and
  Skills (2007), 20–21
Dignity of animals, 101–115
Dignity of creatures, Swiss origin, 103
Direct payments, conditions, 191
Doha round of negotiations, 187
Domestic supply management schemes, 189
"Dumping Alert Mechanism," 194
  elements, 194f
Dumping prices, 6, 191, 195

E
EAA, *see* Ecumenical Advocacy Alliance
  (EAA)
EACTSO, *see* Empresa Asociativa de
  Campesinos de Transformación y
  Servicios Otoreña (EACTSO)
Economic Subsidiarity, principles, 194
Eco-social "regional market economies," 147
Ecumenical Advocacy Alliance (EAA), 4
EFSA, *see* The European Food Safety
  Authority (EFSA)
Eidos, 110–111
EKAH, *see* Swiss Federal Ethics Committee
  on Non-Human Biotechnology
  (EKAH)
Elements, animal integrity, 107
EM, *see* Ethical matrix (EM)
Empirical claims, 49–50
Empresa Asociativa de Campesinos de
  Transformación y Servicios Otoreña
  (EACTSO), 127

"Enabling policy environment," 191
Epistemological values in science
　elegance, 50
　fecundity, 50
　simplicity, 50
Erewhon, 7
Ethanol, 154, 157–159
　production, 157
Ethical matrix (EM), 2, 17–28
Ethical principles, 5, 8, 24–25, 52, 102
Ethics and GM foods
　applied ethics, 49
　ethically justifiable conclusions
　　empirical claims, 49–50
　　normative claims, 50
　ethical responsibilities of scientists, 50–51
　　epistemological values, 50
　　personal values, 50
　　university scientists/industry scientists, challenges faced, 51
　GM foods, production techniques
　　"foreign" genes, use of, 49
　GM technology, ethical issues
　　To Engage in ag Biotech Is Illegitimately to Cross Species Boundaries, 55–56
　　To Engage in ag Biotech Is to Commodify Life, 56–57
　　To Engage in ag Biotech Is to Invent World-Changing Technology, 55
　　To Engage in ag Biotech Is to Play God, 54–55
　　objections, intrinsic/extrinsic, 53
　method for addressing ethical issues, 51–52
　　See also GM foods, method for addressing ethical issues
　minority views, 64
　precautionary approach to new foods
　　ethical dilemma, assumptions to face, 60–61
　participants in willingness-to-pay experiment, treatment analysis/results, *see* Willingness-to-pay experiment
　the precautionary principle
　　GM crops, environmental degradation prevention, 62–63
　　Rio Declaration on Environment and Development (1992), 62
　religion and ethics, 63–64
　　overridingness of ethics, 63
EU directive 2001/18/EC, 88
EU, precautionary principle, 70–72
EU regulation, 70–73
EU responsibility, 70, 77–78
EU, sustainable development, 74–83
Extraterritorial obligations (ETO), 120
Extra-territorial responsibility, principles, 194, 200

F
Fair trade, criteria/goals, 12, 18, 22–23, 42, 144–145, 185, 194–197, 200, 207
Family farming, 6, 190–195
FAO, *see* The Food and Agriculture Organization (FAO)
Fecundity, 50
FIAN, *see* FoodFirst Information and Action Network (FIAN)
Food chain, 1–2, 22, 31–45, 70, 206
FoodFirst Information and Action Network (FIAN), 4, 120
Food irradiation, 58–60
Food production, 2, 10, 20, 22t–23t, 25, 32, 34–36, 38, 42, 47–115, 144, 147, 149, 153, 160, 187, 190, 205–207
Food safety, 32, 33t, 34, 36t, 58, 143, 162, 189, 207
Food security, 4, 6, 36t, 53, 120–121, 126–127, 139–140, 147, 149–151, 162, 187
Food traceability, 2, 31–33, 43–44
Food *vs.* fuel, governance potential for water rivalry 153–160
"Foreign" genes, 49
Frontiers of Justice, 110
Fundamental principle of agriculture, 9
Fundamental principle of modern industry, 9

G
GATS, *see* General Agreement on Trade in Services (GATS)
General Agreement on Trade in Services (GATS), 188
Genetically modified (GM), 2–3, 11, 49–66, 68, 73, 87–97, 106
Genetically modified organisms (GMOs), 3, 33, 53, 57, 64, 67, 70–73, 87–97
　normative concerns, 96–97
　precautionary principle, 88–93
　scientific uncertainties, 91–93
Genetic engineering, 10–11, 54, 56–57, 62, 64, 69, 149–150
Genetic integrity, 113
German Animal Protection Law, 102
　Mitgeschöpflichkeit, basic principle in, 102
Ghana, rice trading/right to food
　food crisis, impact on farmers/women, 124

imported rice, advantages over local rice, 124
import surge, reasons
  low level of world market/CIF prices of imported rice, 122
  policy reasons to boost imports, 123–124
  obligations/responsibilities to right to food, IMF, 124
  rice imports, micro-level study in Dalun, 123–124
Globalization, food ethics, 1–13
Global obligations, 120, 124, 128, 130–133
Global public goods, 170–175
Global water crisis, 154–155, 160
GM, see Genetically modified (GM)
GM foods, method for addressing ethical issues
  consensus and compromise, difference between, 52
  ethical principles for guidance
    Rights theory, 52
    Utilitarian theory, 52
    Virtue theory, 52
  harm envisaged, 51
  information, 51
  moral closure, 52
  options, 51–52
GMO, biosafety research, 87–97
  normative concerns
    prominent mistakes, 96
    solutions to overcome mistakes, 96–97
  precautionary principle and GMO research
    Cartagena Protocol, role in, 88
    hypothesis testing, type I-errors vs. type-II errors, 90–91
    role/responsibility of scientists, 90
    "safe until proven otherwise," assumption, 90
    scientific uncertainties, context, 91–93
    type-II errors and early warnings, 91
  precautionary principle, definitions/applications
    argumentative and prescriptive versions, 89–90
    Bergen Declaration (1990), 89
    Cartagena Protocol, 89
    Rio Declaration on Biosafety, 89
    Wingspread Statement (Raffensperger and Tickner), 89
  scientific dissent, GMO use/release
    adverse effects, 94–95
    cross-disciplinary communication, 96
    "more research," still an uncertainty, 95
    Sarewitz, arguments, 95
  transgenic GMOs, complex interactions by adequate human and environmental models, design of, 94
    ecosystem level, 93
    ethical and social implications, 93
    genomic level, 93
    organismal level, 93
    population level, 93
    small to large-scale release, effects, 94
GMO research and precautionary principle
  Cartagena Protocol, role in, 88
  role/responsibility of scientists, 90
  "safe until proven otherwise," assumption, 90
  scientific context
    nontarget effects, direct/indirect, 92
    uncertainties/ignorance of hazards, 92
    W&H framework, model-based decision support, 92
  type I-errors, hypothesis testing
    choice of H_0, 90–91
    determination/minimization of, 90–91
    risk-related research, 91
  type-II errors and early warnings
    GMO study, controversies, 91
GM "orphan" foods, 60–61
GMOs, see Genetically modified organisms (GMOs)
GM technology, ethical issues
  objections, intrinsic/extrinsic, 53
  UE opposed (Comstock 2000)
    To Engage in ag Biotech Is Illegitimately to Cross Species Boundaries, 55–56
    To Engage in ag Biotech Is to Commodify Life, 56–57
    To Engage in ag Biotech Is to Invent World-Changing Technology, 55
    To Engage in ag Biotech Is to Play God, 54–55
Grameen Danone, 5, 145–146
Grameen Danone Foods Ltd., 145–146
Green Genetic Engineering, 149
Guidelines for Research Involving r-DNA Molecules, 68

H

HACCP, see Hazard Analysis of Critical Control Points (HACCP)
Hazard Analysis of Critical Control Points (HACCP), 32

Helmholtz Centre for Environmental Research, Germany, 5
Herbicide-resistant oilseed crops, 92
High sugar, salt and fat foods (HSSFF), 21, 24, 26–27
History of ideas, property and democracy of water, 167–182
   water conflicts and principles of water democracy
      building dams, control of water resources, 176
      climate change, increased risk of water crisis, 179–180
      conflict over Nile, 177
      as conflicts of 'power,' 177
      Global Policy Forum, 175
      impact on economy and livelihood, 177
      Israeli–Palestinian conflict, 177
      large dams constructed, conflicts between nations, 177
      McGee, view, 176
      Nile River Basin Strategic Action Program, 177
      Omosa, case study of Wajir District in Kenya, 177–178
      OXFAM-GB, democratic water management, 178
      principles of Water Democracy, Vandana Shiva, 181
      UN guidelines for water sharing, 177
      Water Conflict Bibliography, 175
      World Commission on Dams, 176
Homo agrarian, 6
Homo sapiens sapiens, 6
Honduran Institute of Agricultural Marketing (IHMA), 125
Honduras, rice trading/right to food, 125–127
Honesty, 50
HSSFF, *see* High sugar, salt and fat foods (HSSFF)
H_0 (the null hypothesis), 90
Human breeding, 108
"Human dignity and bioethics," 102
Human dignity, debate on (US)
   "human dignity and bioethics," 102
   Ruth Macklin, arguments, 102
Human right to food, 6, 119–134, 186–187
Hunger, *see* Hunger, poverty, and climate change, impact on ethical changes
Hunger, poverty, and climate change, impact on ethical changes
   climatic changes, impact on agricultural yields, 138–139

   Australian Meteorological Association, "normal case" scenario prediction, 139
   glacier size variations, fear of water supplies, 138
   man-made developments, impact on monsoons, 139
   precipitation changes, Latin America and Africa, 139
   Global Monitoring Report 2008, 138
   global warming, 138
   Grameen-Danone, a new business solution
      businesses as social enterprises, 145
      Danone and Grameen, joint ventures, 145
      Grameen–Danone Alliance, Bangladesh, 145
      local entrepreneurs, the beneficiaries, 145
   Institutional approaches toward sustainable solutions
      binding international standards, animal/plant health sectors, 140–141
      civil disobedience, 148–151
      conservation agriculture with its zero tillage, effects, 143
      FAO, aim/goals, 139–140
      higher humus content of soil, effects/benefits, 143
      organic agriculture, international guidelines, 143
      organic agriculture practices, World Bank recommendations, 143
      subsidiarity and solidarity as social ethics principle, 146–148
      voting rules, 142
      voting sessions/composition of delegation, participation of states in, 142–145
      See also Institutional approaches
   IPCC report, 138
   MDG, role, 137–138
   World Food Summit in 1996, goal, 137
Hypertension, 20

I
ICESCR, *see* International Covenant on Economic, Social, and Cultural Rights (ICESCR)
IFOAM, *see* International Federation of Organic Agriculture Movements (IFOAM)

IFOAM, basic principles for worldwide
  organic agriculture
    principle of care, 143
    principle of ecology, 143
    principle of fairness, 143
    principle of health, 143
IgNobel Peace Prize 2008, citizens of
    Switzerland
  'plants have dignity,' concept of, 101
IGOs, *see* Intergovernmental organizations
    (IGOs)
IHMA, *see* Honduran Institute of Agricultural
    Marketing (IHMA)
Indonesia, rice trading/right to food, 128–131
Industry scientists, 51
Ingensiep, H. W., 1–13
Inherent value, concept, 108–109
Innovative and entrepreneurial skills, 27
Institutional approaches
  civil disobedience, 148–151
  FAO, aim/goals
    combat environmental/human pesticide
      poisoning, 139–140
    combat hunger, 139
    "conservation agriculture," 140
  international fair trade, organization of
    basic labor standards, ILO, 144–145
    "Declaration on basic principles and
      rights at work" (1998), 144
    "New World Trading Order," support
      to, 144
  organic agriculture practices, World Bank
    recommendations
    EU-defined standards, 144
    obstacles, 143–144
  subsidiarity and solidarity as social ethics
    principle, 146–148
  voting rules, 142
  voting sessions/composition of delegation,
    participation of states in
    worldwide organic agriculture, basic
      principles (IFOAM), 143
Integrated Pest Management, aim, 140
  FAO, Code of Conduct on application of
    chemicals, 140
  plant disease reduction, 140
"Integrity," 4, 102–103, 107–109, 113,
  187–188, 197, 202
Intergovernmental organizations (IGOs),
  4, 121
International biofuels policy, 160–163
International commodity trade, role, 155

International Covenant on Economic, Social,
  and Cultural Rights (ICESCR),
  119–120
International Federation of Organic
  Agriculture Movements (IFOAM),
  141, 143–144, 197
International Organisation for Standardisation
  (ISO 9000), 32
International Virtual Water Trading Council
  (McKay 2001), 162
Intrinsic value, 4, 18t, 109–110, 112
IPPC, *see* The International Plant Protection
  Convention (IPPC)
ISO 9000, *see* International Organisation for
  Standardisation (ISO 9000)
Israeli–Palestinian conflict, 177

J
Jatropha, 157, 160
John Locke, 5, 169–171, 176
Justice, 2, 12–13, 19, 27, 51, 64, 110, 150, 172,
  202, 207

K
Knockout-mice, 106
Kunzmann, P., 3–4, 101–115
Kyoto Protocol, 157, 163

L
Labelling schemes
  French Label Rouge, 34
  RSPCA Freedom Food label, 34
  UK Red Tractor scheme, 34
Law for the Modernization of the Agricultural
  Sector (LMA), 126
Liberalization, rice trade, 120–131
Livelihood security, sake of, 187
LMA, *see* Law for the Modernization of the
  Agricultural Sector (LMA)

M
"Manifesto" character, 102
Mansholt Plan, 9
MDG, *see* Millennium Development Goals
  (MDG)
Meinhardt, M., 1–13
Mepham, B., 2, 17–28, 107
Mexican "tortilla crisis" (2007), 5, 153–154
Millennium Development Goals (MDG),
  137–138, 148, 155
Mitgeschöpflichkeit, 102
"Modern" utopians, 7
"Most Favored Nation," 200

Multi-functionality, investment in, 190–191
Myhr, A. I., 3, 87–97

N
NAFTA, *see* North American Free Trade Agreement (NAFTA)
Nano-biotechnology, 73
National Biosafety Commission (NBC), 72
National Dairy Development Board of India, 192
National Health Service (NHS), 20
"National Treatment," 200–202
NBC, *see* National Biosafety Commission (NBC)
Neoliberal theories of private property (John Locke), 5
"New Atlantis" (1627), Francis Bacon, 6
NGOs, *see* Non-governmental organisations (NGOs)
NHS, *see* National Health Service (NHS)
Nile, 177
Nile River Basin Strategic Action Program, 177
Non-governmental organisations (NGOs), 1, 4, 31, 95, 139, 142, 160, 192–193, 206
"Nontariff barriers," concept, 201
"Normal weight," definition (WHO), 20
Normative claims, 50
North American Free Trade Agreement (NAFTA), 188
Norwegian Gene Technology Act (1993), 88, 96
"Novum Organum," Francis Bacon, 167
Nuffield Council on Bioethics 2007, 21
Nutritional research and technology, 36t

O
Obese, definition (WHO), 20
Obesity
 crisis, 2, 17–28
  See also EM, tool in policy interventions of obesity crisis
 definition, 20
 wellbeing/autonomy/fairness, 18t, 20
 POM, 23t
 SPM, 22t
Obesogenic environment, 21, 25
OIE, *see* The World Organization for Epizootics (OIE)
"Open access" elements, 171
Organic agriculture, 73, 75, 141, 143–144, 193, 197–198
Overweight, definition (WHO), 20

Ownership rights, categories
 common ownership, 171
 free access, 171
 private ownership, 171
 state ownership, 171

P
Partzsch, L., 5, 153–164
Pastoralism, 177–178
Performance-based regulation, 26
Personal responsiveness, 80
Personal values in science
 honesty, 50
 responsibility, 50
Plant breeding, 49
Policy decisions, 19, 21–25, 27, 62, 64
Policy framework, 6, 158, 191–193
Policy objectives matrix (POM), 22, 23t
"Political space," 186
POM, *see* Policy objectives matrix (POM)
Post-WTO architecture of agricultural trade, 200–202
Poverty, 4, 122–123, 125, 131, 137–151, 155, 162, 196
 See also Hunger, poverty, and climate change, impact on ethical changes
Precaution and sustainable development, agro-food biotechnology
 dimensions of a precautionary attitude
  collective engagement, 79–80
  continuous learning process, 82–83
  goal-oriented approach, 78–79
  projections of the future, 81–82
 precaution lost its orientation
  Bergen definition of precautionary principle states, 74
  harm-oriented approach, 74–75
  Jensen's arguments, 74
  technological applications, step stone to precaution, 75
 re-linkage
  German concept of *Vorsorge*, 75
  goal-oriented interpretation, 75–76
  total, continuous, and future-oriented responsibility, 77
Precautionary attitude, dimensions of
 collective engagement, 79–80
 continuous learning process, 82–83
 goal-oriented approach, 78–79
 projections of the future, 81–82
Precautionary principle, 3, 61–63, 67, 70–75, 78, 83, 87–91, 96
 prescriptive versions

Cartagena Protocol, 89
  environmental protection, cost-effective nature of, 89
  environmental risk, cost-benefit analyses of, 89
  Rio Declaration on Biosafety, 89
Precautionary principle states, Bergen's definition, 74
Principles of Water Democracy, Vandana Shiva, 181
Production history, 32, 33t, 34–35, 36t, 37f, 38–43
Property, 5, 7–8, 52, 148, 167–182, 201
  See also History of ideas, property and democracy of water
"Protection beyond pain and injury," 105
Public debates on GM crops, historical reconstruction
  potential environmental risks, severity/acceptability
    biotechnology/multinational companies, socio-economic issues, 69
    food quality issues, focus, 69
    intensive farming and biodiversity, relation, 69
    legal coexistence frames, issue, 69
  r-DNA technique, patent issue (1976)
    Bayh-Dole Act (1980), adoption, 68
  r-DNA technique, safety issues (1970)
    Guidelines for Research Involving r-DNA Molecules, 68
Public duties/public goods, 172–173
  climate, global public good, 172
  global public goods, Kaul
    goods identified in vision of international community, 174
    sustainability, 175
    water, global natural "commons," 175
  International cooperation in the twenty-first century, 173
  liberalization and privatization of, 172–173
  "Providing Global Public Goods: Managing Globalization" (2003), 173
    public–private partnership concept, 174
  public and the private spheres, reconfiguration
    political approach, need for, 175
  theory of public goods, Keynesianism period, 174

Q
Qualified market access, 198–200, 201f

R
Rawlsian "overlapping consensus," 25
r-DNA technique, 68
Reality and utopia, justice as principle, 12–13
  rational background theory of justice, Ekardt, 12
    Consensus Conferences, 13
    reasoned arguments, 12
Regional markets, 6, 147, 192
Regulatory reaction, 3, 70–73
Religious arguments, 64
Resource-Conserving Agriculture, 193
Rice trade policies, 4, 121
Rights theory, 52
Rio Declaration on Biosafety, 89
Roundtable on Sustainable Palm Oil (RSPO), 163
3R principle, 106
RSPCA, see UK Royal Society for the Protection of Animals' (RSPCA)
RSPO, see Roundtable on Sustainable Palm Oil (RSPO)

S
Sachs, W., 5–6, 185–202
"Salomon's House," 6–7
Sanitary and Phytosanitary Measures (SPS), 141–142, 188
SAPs, see Structural Adjustment Programmes (SAPs)
SBB, see The Section on Biosafety and Biotechnology (SBB)
Scientific responsibility, 50–51
Setting standards, quality trade, 196–197
"Sexual Contract" (1988), 169
Simplicity, 50
Social ethics principle, subsidiarity/solidarity, 146–148
  eco-social "regional market economies," 147
  food equality/regional self-determination, 147
  micro-finance and micro-insurance, 148
  regional farming, processing, and merchandising, cooperation with, 148
  self-help measures for trading/opening new markets, 148
Society members, 27
Solidarity, 5, 9, 146–148, 180, 200
  See also Subsidiarity/solidarity as social ethics principle
Sound argument, meaning, 49

Special Safeguard Mechanism, 187
Species, 110–111
Species-specificity, 108
Specified principles matrix (SPM), 22, 22t, 24–25
Speech act theory, 102–103
SPM, *see* Specified principles matrix (SPM)
SPS, *see* Sanitary and Phytosanitary Measures (SPS)
Structural Adjustment Programmes (SAPs), 121
Subsidiarity, 5, 146–148, 181
Support without dumping, 193–195
"Survival of the fittest" principle, Spencer/Darwin, 8
Sustainability
  process/production standards, 197–198
  sake of, 188–189
  water systems, 154, 161
Sustainable Development Commission 2008, 21
Sustainable family farming, frameworks, 191–193
"Sustainable Production and Consumption," Germany, 5
Swiss Federal Ethics Committee on Non-Human Biotechnology (EKAH), 101, 104, 107, 113–114
Synthetic biotechnology, 73

T

TBT, Technical Barriers to Trade (TBT)
Technical Barriers to Trade (TBT), 188
Telos, 111–112
"*The Developmental Dictionary: A guide to knowledge as power*", 6
The European Food Safety Authority (EFSA), 70
The Food and Agriculture Organization (FAO), 120–122, 125–128, 133, 139–143, 148, 199
The International Plant Protection Convention (IPPC), 141–142
Theodore Roosevelt (1901–1909), 176
The Schweisfurth foundation – a German food-ethics-platform, 205–207
The Section on Biosafety and Biotechnology (SBB), 73
The World Organization for Epizootics (OIE), 141
Traceability *vs.* ethical traceability, 36
Trade liberalization, 120–122, 132–133, 185
Trade-Related Investment Measures (TRIMS), 188

"Tragedy of the Commons," essay, 170–171
TransFair seal, 145
Transgenic organisms, "living machines," 10
Transparency, *see* Ethical traceability, food chain transparency
TRIMS, *see* Trade-Related Investment Measures (TRIMS)
Type 2 diabetes, 20

U

UE, *see* Unnatural to genetically engineer plants, animals, and foods (*UE*)
UK Royal Society for the Protection of Animals' (RSPCA), 34
UN guidelines for water sharing
  no harm rule, 177
  rule of equitable and reasonable use, 177
University scientists, 51
Unnatural to genetically engineer plants, animals, and foods (*UE*), 53
"Upstream" and "downstream" innovation process, 96
Utilitarian theory, 52
Utopia, 1–13
'Utopian dreamers,' 9

V

Virtual water, 5, 154–164
  See also Virtual water, rethinking a resource
Virtual water accounting, 5, 154, 156, 159f, 162–163
Virtual water trade, 158–160
Virtue theory, 3, 52
von Winterfeld, U., 5, 167–182

W

Walker & Harremöes (W&H) framework, 92–93
Wal-Mart, 33
Water Conflict Bibliography, 175
Water conflicts, *see* Water conflicts and principles of water democracy
Water conflicts and principles of water democracy
  building dams, control of water resources, 176
  climate change, increased risk of water crisis, 179–180
  conflict over Nile, 177
  as conflicts of 'power,' 177
  Global Policy Forum, 175
  impact on economy and livelihood, 177
  Israeli–Palestinian conflict, 177

large dams constructed, conflicts between nations, 177
McGee, view, 176
Nile River Basin Strategic Action Program, 177
Omosa, case study of Wajir District in Kenya, 177–178
OXFAM-GB, democratic water management, 178
principles of Water Democracy, Vandana Shiva, 181
UN guidelines for water sharing, 177
Water Conflict Bibliography, 175
World Commission on Dams, 176
"Water footprints," 163
West Java's rice barn, 130
World Commission on Dams, 176

World Summit on Sustainable Development (WSSD), 188
World Trade Organization (WTO), 6, 73, 120–121, 129, 132, 134, 141–142, 147, 161–162, 185–188, 191, 194, 200–202
World Wide Fund for Nature (WWF), 163
WSSD, *see* World Summit on Sustainable Development (WSSD)
WTO, *see* World Trade Organization (WTO)
Würde, 102
WWF, *see* World Wide Fund for Nature (WWF)

Z
Zoocentric, 108

Printed by Books on Demand, Germany